Monographs on Astronomical Subjects: 6

General Editor, A. J. Meadows, D.Phil.,

Professor of Astronomy, University of Leicester

Cosmic X-ray Astronomy

In the same series

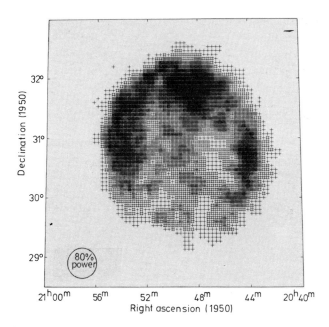

Frontispiece. An x-ray picture of the Cygnus Loop supernova remnant obtained with a focusing x-ray telescope on a sounding rocket. From Rappaport *et al* (1979).

Cosmic X-ray Astronomy

D. J. Adams

University of Leicester

Monographs on Astronomical Subjects: 6

Adam Hilger Ltd, Bristol

British Library Cataloguing in Publication Data

Adams, D J
    Cosmic X-ray astronomy.—(Monographs on
    astronomical subjects; 6 ISSN 0141-1128).
    1. X-ray astronomy
    I. Title   II. Series
    522′.6   QB472
    ISBN 0-85274-253-3

Published by Adam Hilger Ltd, Techno House, Redcliffe Way, Bristol BS1 6NX
The Adam Hilger book-publishing imprint is owned by The Institute of Physics

Typeset by The Universities Press (Belfast) Ltd, and printed in Great Britain by The Pitman Press, Lower Bristol Road, Bath

Preface

This monograph reviews the progress made in x-ray astronomy over the period 1962 to 1978. It is written for the non-specialist with some knowledge of astronomy and physics. The launch of the Einstein observatory satellite late in 1978 was such a significant step in the development of the subject, so likely to lead to an enormous increase in our knowledge of it, that the present moment seems eminently suitable for a review of existing knowledge.

The Einstein observatory, or HEAO-B, as it used to be known, is a focusing x-ray telescope with imaging detectors. It is capable of obtaining proper x-ray images at high resolution of selected astronomical objects in the energy range 0·25–4 keV. At the time of writing, some of its early observations are being made known in the form of preprint material and popular magazine articles. These articles are based on the television screen pictures which have been made available so far. One can foresee the appearance of several bumper editions of the *Astrophysical Journal* once these data have been properly digested.

Einstein has been pointed towards a number of Galactic supernova remnants. Cassiopeia A, for example, shows very much the sort of structure which was expected. It is a shell-like source, approximately circular in appearance, and brighter around the edges. Local regions show enhanced emission, rather as in the radio part of the spectrum. The overall impression is of a very much scaled down version of the Cygnus Loop, as seen in the frontispiece. Observations of other Galactic fields have revealed faint x-ray emission from hot stars. Einstein is several hundred times more sensitive to faint star-like sources than were satellites of the Uhuru type.

It is in the field of extragalactic astronomy that Einstein has made its greatest impact. The normal spiral galaxy M31,

vii

which had previously been only just detected, is shown to contain fifty or more star-like sources. Many are concentrated around the centre of the galaxy, and have luminosities similar to the bright x-ray stars in our own Galaxy. Centaurus A, the nearby radio galaxy, shows a point-like nuclear source, together with diffuse emission to the north-east. The Virgo cluster of galaxies shows, rather surprisingly, x-ray emission concentrated around the major galaxies in the cluster, and very little in between. Other clusters show more uniform diffuse emission. These results are only a foretaste of what is to come from Einstein.

Thanks are due to Dr R. S. Warwick and to other members of the Leicester x-ray astronomy group for many helpful discussions.

D. J. Adams
Leicester University
July 1979

Contents

1. General Observational Features

1.1. Introduction

Cosmic x-ray sources were discovered by accident in 1962, when the American Science and Engineering group of the United States launched a rocket with sensitive detectors to search for x-radiation from the Moon. This experiment was made as a small part of the preliminary investigations for the Apollo manned spaceflight programme. From the point of view of lunar research, the investigation was disappointing, and no signal was detected. However, the instrument scanned a fairly wide region of sky, and detected x-radiation from the direction of the constellations Scorpio and Sagittarius. This result was published by Giacconi *et al* in 1962. Astronomical x-radiation had previously been detected from the Sun, but it soon became clear that the x-rays detected in 1962 came from distances much greater than the size of the Solar System.

This detection of x-rays from distant astronomical objects was quite unexpected. The nearest star lies some 100 000 times further away from us than the Sun. The x-radiation from the Sun amounts to a very small fraction of its total energy output, and a simple calculation shows that, even if the nearest star emitted x-rays at one hundred times the rate

Note: CGS units have been used throughout this monograph because that is the prevailing practice in x-ray astronomy, and in astrophysics in general. The following units are used in the text without reference to their SI equivalents:

$$1 \text{ cm} = 10^{-2} \text{ m}$$
$$1 \text{ g} = 10^{-3} \text{ kg}$$
$$1 \text{ erg} = 10^{-7} \text{ joules}$$
$$1 \text{ erg s}^{-1} = 10^{-7} \text{ W}.$$

of the Sun, it would have been quite undetectable by the x-ray telescopes of the 1960s. It is now known that many x-ray sources are very much more distant than the nearest star: indeed, many of the bright ones lie 10 000 times further away. The conclusion must be that these x-ray sources are intensely luminous. How they come to shine so brightly therefore becomes an important question in astrophysics.

The distribution of x-ray sources over the sky was investigated by further rocket observations in the years following the initial discovery. It was found that they are distributed in a band across the sky, coincident with the Milky Way, though the strongest source, Scorpio X-1, lies some 20° above the plane of the Milky Way.

More recent work on the distribution of cosmic x-ray sources, using much more sensitive satellite telescopes, has confirmed these observations. The brighter sources lie in the Milky Way, but fainter sources, which were not seen on the earlier rocket surveys, are distributed more uniformly across the sky. The brighter sources are therefore referred to as Galactic, whereas the fainter, more uniformly distributed, sources are generally regarded as being extragalactic. There is independent evidence to identify a number of the Galactic plane objects with known stars and nebulae in the Galaxy; whereas some of the weaker sources out of the Galactic plane have been identified with known galaxies.

In the following sections, we will briefly review current knowledge of what lies within and beyond our Galaxy in order to place the x-ray sources into perspective.

1.2. Our Galaxy

The Galaxy is a flattened disc with a bulge at the centre, and is made up of stars, rarefied gas and micrometre-sized solid particles (called dust). It has a diameter of about 30 000 parsecs, but a thickness outside the bulge of only 300–500 parsecs. (Here 1 parsec (pc) $= 10^{18}$ cm $= 2 \times 10^5$ times the distance from the Earth to the Sun.) The Galaxy contains about 10^{11} stars; these are more closely packed towards the centre than at the periphery of the disc.

On a moonless night, the flattened structure of the Galaxy

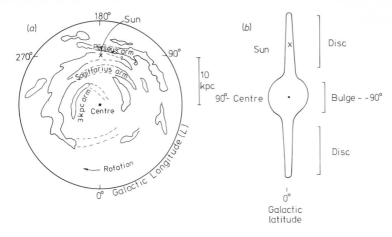

Figure 1.1. Schematic plan and elevation of our Galaxy.

shows up as a bright band of faint stars—the Milky Way. The bright stars, on the other hand, appear to be more randomly distributed across the sky. This is because they lie at distances which are smaller than the thickness of the disc. Visible-wavelength observations of the Milky Way have to be treated with some caution because the dust in the plane of the Galaxy limits visibility to a few thousand parsecs at best. For this reason the centre of the Galaxy is completely obscured in visible light. The dust does not, however, impede the view of radio and infrared astronomers: the best observations of the structure of the Galactic disc have been made by radio astronomers using the 21 cm spectral line of neutral atomic hydrogen. It is mainly from 21 cm work that the spiral structure of our Galaxy has been traced. This spiral structure is a feature which can be more readily discerned in photographs of some external galaxies.

The disc of the Galaxy is in rotation about an axis through its centre, normal to its plane. The stars are in roughly circular orbits about the centre of the Galaxy, and move with an angular velocity which decreases with distance from it. Stars at the distance of the Sun (about two-thirds of the way out) from the centre take about 10^8 years to complete an orbit. The age of the Galaxy is about 10^{10} years.

Not all of the stars in the Galaxy are as old as 10^{10} years. The so-called Population I stars found in the spiral arms are thought to have been formed more recently, and many of them will burn out their nuclear fuel in a period of less than 10^{10} years. Indeed, there is evidence that stars are still forming out of collapsing clouds of interstellar gas. It is the oldest stars, known as Population II, that are probably about 10^{10} years old. They are not confined to the disc of the Galaxy, but are found both singly and clustered together in dense spherical associations (globular clusters) in the halo of the Galaxy.

1.3. Galactic X-ray Sources

In this section the general properties of the Galactic x-ray sources will be considered, namely, their distribution in the sky, their distances, their luminosities, their spectra, their variability, and their identification with objects known to optical and radio astronomers. A discussion of the individual sources is deferred to Chapters 4 and 5.

The distribution of the sources in the sky has already been mentioned, and can be seen in figure 1.2(a). The brighter sources trace out the position of the plane of the Galaxy which is inclined at about 60° to the equator. The brighter Galactic x-ray sources are dispersed across the Galactic equator with a mean deviation of about 3°, except for Sco X-1 and Her X-1, which lie 20° and 35° off the plane, respectively. Taking this figure of 3°, it is possible to estimate the mean distance of the sources using the following argument. Suppose that the sources are dispersed about the central plane of the Galaxy by a linear distance of 200 pc. This dispersion will be the product of their mean distance, $\bar{d}$, and the tangent of their angular dispersion. Thus

$$\bar{d} = \frac{300}{\tan 3°} \, pc = 4 \, kpc = 4000 \, pc.$$

It is unnecessary to argue about the precision of the numbers taken and the precise meaning of the averages used, to see that the majority of the x-ray sources must be at considerable distances from us in the Galaxy. Sco X-1 and Her X-1, which were excluded from the above argument, are thought to lie within 1 kpc distance of the Sun.

4

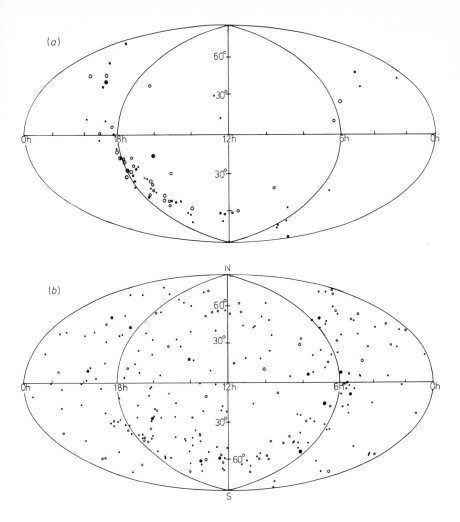

Figure 1.2. (*a*) The bright x-ray sources plotted in celestial coordinates (right ascension and declination). For a map plotted in Galactic coordinates, see figure 4.12. (*b*) A map of bright visible stars for comparison with (*a*).

The second feature which is apparent in the map of source positions is the tendency of the sources to cluster around the direction towards the centre of the Galaxy and, to a lesser extent, around the Cygnus region. The Galactic centre clustering suggests the likelihood of a real concentration towards the centre of the Galaxy. The clustering in the Cygnus region may indicate a concentration of sources in the local spiral arm where the Sun is situated, since this spiral arm runs

5

through the Cygnus and Orion directions in the sky. The latitude dispersion around $l_{II} = 0$ is smaller than that in Cygnus, suggesting that the low Galactic longitude sources are more distant.

The luminosity of an x-ray source can be estimated from the observed x-ray flux, once the distance is known. To arrive at a sample luminosity, a source at the distance of the Galactic centre will be considered. Typically, the brighter sources in the Sagittarius region of the sky are detected at the level of 1 x-ray photon $cm^{-2} s^{-1}$. Each x-ray photon is characterised by an energy in the 1 keV ($= 1.6 \times 10^{-9}$ erg) range. We will adopt 2·5 keV as the average energy. The energy flux falling on the top of the Earth's atmosphere then amounts to some 4×10^{-9} erg $cm^{-2} s^{-1}$. This flux, F, is related to the luminosity, L, of the source and the distance, D, by

$$F = \frac{L}{4\pi D^2}.$$

1 pc $= 3 \times 10^{18}$ cm, so $d = 3 \times 10^{22}$ cm and $L = 5 \times 10^{37}$ erg s^{-1}. This may be compared with the total luminosity of the Sun, which is 3.8×10^{33} erg s^{-1}. Thus a typical x-ray source is more than 10 000 times more luminous in x-rays than the total radiative output of the Sun. Admittedly the x-ray sources close to $l_{II} = 0$ are amongst the most luminous in the Galaxy, but even the closer sources, such as Her X-1 and Sco X-1, have luminosities in the 10^{35}–10^{36} erg s^{-1} range. It is this feature which makes them astrophysically important. An independent confirmation of the high luminosity of Galactic x-ray sources comes from the observation of a similar source in the Small Magellanic Cloud (SMC X-1). In this case, the distance is known fairly accurately, and the luminosity calculated, taking spectral details into account, amounts to 10^{38} erg s^{-1}.

So far, we have established that Galactic x-ray sources are distant, and therefore populate the Galaxy rather sparsely. On the other hand, the individual sources are highly luminous. Before discussing in more detail the physical nature of these sources, it is worthwhile considering their total luminosity over the whole Galaxy. This figure can then be compared with the integrated Galactic luminosity in the visible region, which encompasses most of the emission of normal-type stars. We will suppose that the Galaxy contains 200 x-ray sources with a mean luminosity of 10^{37} ergs^{-1}, giving a total luminosity for the Galaxy of 2×10^{39} ergs^{-1}. There are in the Galaxy about 10^{11} visible stars with a mean luminosity of, say, 10^{33} ergs^{-1}, giving a total visible luminosity of about 10^{44} ergs^{-1}. Thus, taking the Galaxy as a whole, its x-ray emission accounts for only a small fraction of the total energy emission. The lifetime of most x-ray sources is not known, but it is likely that they represent brief luminous phases in the life cycles of some stars.

The identification of Galactic x-ray sources with objects known to visible and radio astronomers is of utmost importance to an understanding of the nature of these objects. Once an optical spectrum is obtained it is possible to work out the composition, the temperature, the density and the radial velocity of the object in many cases, and it is usually possible from this information to classify the object responsible for the emission. At the time of writing, only a fraction of the Galactic x-ray sources have been so identified. The first identification made was of Tau X-1 with the Crab Nebula supernova remnant. This was established in 1964 when the Moon occulted (passed in front of) the nebula. A rocket carrying x-ray sensors was launched in time to observe the disappearance of the Crab behind the limb of the Moon. The x-ray signal was extinguished at just the time when the Crab passed behind the limb. Subsequent work has amply confirmed this identification, and several other supernova remnants are now known to be x-ray sources. A general feature of these x-ray sources is that they are extended emission objects, covering areas of the sky several arc minutes or more across. Two emission mechanisms appear to be involved in the x-ray output of supernova remnants: the thermal emission of a hot gas and the synchrotron radiation of high-energy electrons spiralling in a magnetic field. The synchrotron mechanism is the same process as that responsible for the radio emission of supernova remnants.

The second x-ray source on which attention was concentrated was Sco X-1. Being the brightest source in the sky, this object lent itself to observation by a simple form of

7

modulation collimator (discussed in Chapter 2). The results showed that the angular size of the object was small, and probably star-like. The position was measured to an accuracy which could not be achieved by the normal scanning method. From this positional measurement, and using spectral arguments which are outlined in Chapter 4, Sco X-1 was identified with a faint (13th magnitude) star. A number of other x-ray sources have since been identified with stars, although the methods used have been rather indirect. The basic problem is that the positional measurements are normally only accurate to an arc minute or so, yielding an error box which contains a large number of faint stars. In order to identify an x-ray source on the basis of position alone, it is necessary to make a special measurement accurate to a few arc seconds using a particular technique, such as the timing of an occultation by the Moon. A further problem is that visible light is severely absorbed by the dust in the Galactic disc, so the object may be invisible at optical wavelengths. Even an accurate x-ray position does not therefore guarantee an identification: this difficulty is most severe for sources around the Galactic centre.

The identification of several x-ray sources with visible stars actually derived from the discovery that the x-ray output of these objects turns itself on and off periodically. This discovery was delayed until the Uhuru satellite was operational, since it required the ability to monitor selected sources for extended periods. Cen X-3 was the first source found to turn on and off in a regular fashion, and it has a period of 2·2 days. This periodic behaviour is interpreted in terms of a close binary star system; the disappearance of the x-ray signal is caused when the x-ray component of the binary is eclipsed by the other star. Once a visible star can be found which has an eclipsing behaviour with the same period, the identification is made. A dozen or so of these eclipsing binaries are now known, and they are described in detail in Chapter 4. Their optical counterparts are often found to be hot supergiant stars, which are intrinsically much brighter in the visible than the optical counterpart of Sco X-1. The x-ray source itself is associated with a very small and dense star, probably a neutron star, which cannot be seen in the visible.

All the x-ray sources in the Galaxy, apart from supernova remnants, are of no more than stellar dimensions. It is thought that most, if not all, are associated with binary systems, even through only a fraction of them exhibit the eclipsing binary behaviour mentioned above. A common feature of x-ray stars is that their x-ray brightness varies on timescales of an hour or less; some even vary over a small fraction of a second. Such rapid variability is very uncommon amongst visible stars, which only appear to twinkle as a result of motions in the Earth's atmosphere.

X-ray stars tend to be classified according to (i) position in the Milky Way, and (ii) prominent brightness changes. The eclipsing binaries have already been mentioned. X-ray transients, such as novae, suddenly flare up, then gradually die away again over a few months. X-ray bursters emit intense and repeated bursts of only a few seconds' duration each. Globular cluster x-ray sources lie in the dense spherical star clusters of that name. Galactic bulge x-ray sources are those which populate the inner reaches of the Galaxy; they seem to differ slightly from the x-ray stars in the spiral arms.

1.4. External Galaxies

On a larger scale, the Universe appears to be made up mainly of galaxies resembling our own. The nearest of these galaxies is the Large Magellanic Cloud at a distance of some 50 000 pc. The well known Andromeda Nebula (M31) lies about 700 000 pc distant, and shows a spiral structure very similar to our own Galaxy. Our own Galaxy and M31 are rather larger than average, whilst the Magellanic Clouds are smaller than average. When the spatial distribution of external galaxies is studied, a strong tendency towards clustering becomes apparent. The 'Local Group' to which our own Galaxy belongs contains about 25 members. The Virgo cluster, on the other hand, lies at a distance of about 10 Mpc (= 10 000 000 pc), and contains some 2500 member galaxies. The Virgo cluster shows up clearly on deep-sky photographs, such as the plates of the Palomar Sky Survey. These external galaxies are only seen at visible wavelengths in directions away from the Milky Way. This latter 'Zone of Avoidance' is a result of the obscuration by the dust in the disc of our own Galaxy.

Galaxies are normally classified into spiral (like M31), barred spiral, elliptical, or irregular (like the Large Magellanic Cloud). This classification is based on their visible appearance, due allowance being made for their inclination to the line of sight. A small fraction of galaxies exhibit rather unusual properties. Radio galaxies show strong radio emission, frequently from two regions equally spaced about the position of the visible galaxy. Seyfert galaxies show signs of an explosive expansion, as indicated by the Doppler-broadened spectral lines emitted from their nuclei, and are also characterised by powerful emission at infrared wavelengths. Quasars, or quasi-stellar objects (QSOs), are star-like in appearance, but are thought to be very distant galaxies emitting ultra-powerfully in all spectral ranges. All these objects probably form a sub-set of a larger family of 'active galaxies'.

The distant galaxies appear to be receding from us. This conclusion is based on the observation that the spectral features in distant galaxies are shifted toward the red. Interpreted as a Doppler shift effect, this implies a velocity of recession. Hubble's law relates the velocity of recession

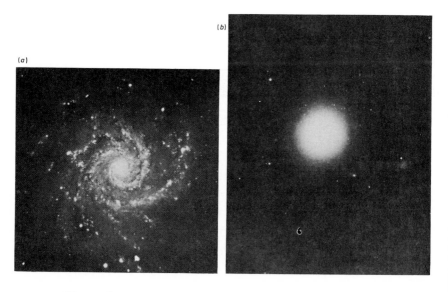

Figure 1.3. Typical (a) spiral and (b) elliptical galaxies.

linearly to the distance, the constant involved being approximately 50 km s^{-1} recession velocity per Mpc distance. This law can only be relied upon for galaxies with distances greater than a few Mpc; at smaller distances the random velocities of the galaxies relative to each other dominate. It is on the basis of these observations that the whole Universe is thought to be expanding, the prevailing view of cosmologists being that this results from an initial 'big bang'.

1.5. Extragalactic X-ray Sources

Extragalactic objects are not prominent in the visible sky. Only the Large and Small Magellanic Clouds and the Andromeda Nebula (M31) are visible to the naked eye. Other galaxies are fainter and require telescopes for their detection. By analogy, one might therefore expect the x-radiation from galaxies to be faint. In the case of M31, this expectation is borne out. It was detected at a low level by the sensitive Uhuru satellite survey, and its x-ray luminosity amounts to about twice that of our own Galaxy. As it is thought to contain about twice as much mass as our Galaxy, this result means that M31 has x-ray properties very similar to those of our Galaxy. Observations by the imaging telescope HEAO-B will be required before the individual sources in M31 can be resolved. The Magellanic Clouds, on the other hand, are more than ten times closer than M31, and it has been possible to pick out individual sources in these rather under-massive galaxies.

It follows from this that the discovery of many x-ray sources at high Galactic latitudes is rather surprising. Some of these sources, however, have been identified with known external galaxies. As an example, we might consider Centaurus A (NGC 5128), a galaxy at a distance of some 4 Mpc, and so lying about six times further away than M31. If it had an x-ray luminosity similar to that of M31, a simple application of the inverse square law shows that it should appear 36 times fainter, at which level it would not yet have been detected. Yet the Uhuru survey showed it to be three times brighter than M31, implying a luminosity some 100 times greater than M31. Having already remarked that individual

Galactic sources are themselves very luminous, we now find that some extragalactic sources are overluminous compared with the total radiation of a galaxy like our own.

One feature stands out: of the identified extragalactic sources, most are either galaxies which show some other signs of abnormal activity, or are the centres of rich clusters of galaxies. The activity in the case of Cen A is evidenced by strong radio emission from regions on either side of the galaxy. In fact, Cen A is the nearest of the 'radio galaxies' and, as its name implies, it is the strongest radio source in Centaurus. As another example, NGC 4151 is an x-ray source, and also shows Doppler-broadened visible emission lines from its nucleus. These broad emission lines are characteristic of Seyfert galaxies, and indicate the occurrence of a violent explosion in the nucleus. Finally, 3C 273 is an x-ray source and a quasar—a visible, infrared and radio source which, on the basis of the large redshift of its visible spectrum, is believed to be at a distance of about 1000 Mpc, and to be one of the most luminous of known astronomical objects.

Other identifications of extragalactic x-ray sources have been with the centres of rich clusters of galaxies. Thus, in the centre of the Coma Berenices cluster there is an extended x-ray source about $0 \cdot 5°$ across. The x-ray sources are found to be similarly extended in the Virgo and Perseus clusters. Some confusion arises in the case of these latter two, however, since the active galaxies Virgo A (M87) and Perseus A (NGC 1275) lie at the respective centres of these clusters. It seems that both of these active galaxies are strong x-ray emitters, whilst there is also x-ray emission from the appropriate cluster centre.

1.6. The Isotropic X-ray Background

The whole sky appears bright in x-rays. This is in strong contrast to the situation in the visible, where the night sky is dark. At x-ray energies above about 2 keV, this background radiation appears to be the same in all directions. It has a spectrum which extends well beyond the x-ray region into the gamma-ray part of the spectrum.

That this background is extragalactic in origin seems fairly clear from its isotropy. Precisely what astrophysical processes are producing it is much less clear. It could be the effect of strong emission from millions of distant galaxies which have not yet been individually resolved, or it could be caused by thermal or cosmic ray processes operating in intergalactic space.

1.7. Names of X-ray Sources

The original system for naming x-ray sources was to call the brightest source in a constellation 'Constellation X-1' or 'Constellation XR-1', the second brightest source 'Constellation X-2', and so forth. Examples are Cyg X-1, Cyg X-2, Cyg X-3, Sco X-1, Sco X-2, where Cyg is the normal abbreviation for Cygnus and Sco is an abbreviation for Scorpio. There are some apparent anomalies in the scheme, owing both to misinterpretations of early data and to transient sources which have disappeared. For example, the brightest source in Centaurus is now Cen X-3. This scheme is still in use for the most prominent x-ray sources.

Since the launch of Uhuru, various catalogues have been produced, and x-ray sources are known by catalogue names such as 3U 1207 +39, A 0620 −00, MX 0513 −40. These names can be interpreted quite simply: 3U 1207 +39 is a source in the third Uhuru catalogue at a position right ascension $12^h 07^m$; declination $+39°$. A 0620 −00 is a source discovered by the Ariel V workers at a position $RA = 6^h 20^m$; $Dec = −00°$ (just south of the equator). MX 0513 −40 is a source first reported by Massachusetts Institute of Technology (x-ray) workers at a position $RA = 5^h 13^m$; $Dec = −40°$. A source may have several different catalogue names, and the catalogue name usually gives no clue as to the nature of the source.

1.8. Quantitative X-ray Astronomy

The remainder of this book will deal mainly with the measurement and interpretation of astronomical x-ray fluxes, luminosities and spectra. It is necessary, therefore, to explain at this stage the rather mixed system of units which is used to express x-ray quantities.

It should be pointed out to the reader who is unfamiliar with astrophysics, that the values quoted are generally of lower accuracy than is usual in most branches of physical science. This applies particularly to distances and luminosities: distances are rarely known to better than 30% accuracy.

1.8.1. X-rays as Electromagnetic Radiation

Most measurements in x-ray astronomy have been made in the energy range 2–20 keV. In a laboratory environment, these would be termed soft x-rays, and have a much smaller penetrating power than the higher-energy (about 100 keV) radiation used to 'x-ray' human bodies and pieces of machinery. Astronomers normally term radiation with energies in the range 20–200 keV 'hard x-rays', whilst the 0·2–2 keV region of the spectrum is called 'soft x-radiation'. The divisions at about 2 and 20 keV arise from the types of detectors used. Figure 1.4(a) summarises the relationship of x-rays to other parts of the electromagnetic spectrum.

In common with visible light, radio waves and other classes of electromagnetic radiation, x-rays exhibit both wave- and particle-like properties. In the case of x-rays, the particle-like properties are more evident, and the concept of x-ray photons is essential. X-radiation of wavelength λ has a frequency $\nu = c/\lambda$ (c is the velocity of light), and is composed of discrete photons of energy, $E = h\nu = hc/\lambda$ (h is Planck's constant). Photon energy (in keV) is the commonest measure of x-ray class, but frequency in Hz (cycles per second) and wavelength, usually in ångström units ($1 \text{ Å} = 10^{-10} \text{ m} = 10^{-8} \text{ cm}$), are also used.

X-rays in the 2–20 keV range are readily absorbed by materials. The mechanism is predominantly photoelectric absorption, and is equally effective in gases, liquids and solids. Materials of higher atomic number absorb x-rays more efficiently. As a consequence, x-rays are quite unable to penetrate the Earth's atmosphere, which has a column

14

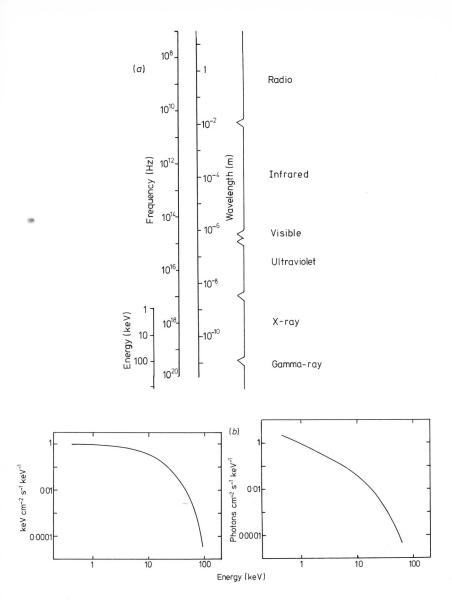

Figure 1.4. (*a*) The electromagnetic spectrum. (*b*) A typical x-ray spectrum plotted in two different standard systems.

density of about $1000\,\mathrm{g\,cm^{-2}}$. Therefore 2–20 keV astronomical observations must be conducted from altitudes of more than 100 km using sounding rockets or satellites. Harder x-rays are more penetrating, and observations can be made from balloon platforms at about 30 km altitude. X-rays of energies greater than about 4 keV are able to traverse the

15

Galaxy, but soft x-radiation with energies below 2 keV suffers severe attenuation in the interstellar gas of the Galaxy.

X-rays are also detected by means of their interaction with matter. Single photons can be observed, but the detectors also respond to some extent to the cosmic radiation. X-ray optics are difficult to construct because neither lenses nor conventional mirrors function in this spectral region. The only way to provide optical elements is to employ solid surfaces as grazing-incidence reflectors, or to use crystal lattices as 'diffraction gratings'. The basic physics of x-radiation is dealt with in most texts on modern physics.

1.8.2. X-ray Fluxes and Spectra

The flux from an astronomical source is the amount of energy received from it at the top of the Earth's atmosphere. It depends upon the distance of the source, as well as upon the intrinsic energy output, or luminosity, of the source.

Photon fluxes are the simplest measurement of an x-ray source flux, with the most straightforward unit being that of photons $cm^{-2} s^{-1}$. Photon count rates are related to photon fluxes, but depend on the effective area and spectral efficiency of the detector used. For example, 'Uhuru counts' are expressed in counts s^{-1}, and represent the count rate detected by the Uhuru satellite detectors (which had an effective area of 840 cm^2, and were sensitive over the approximate energy range 2·5–6·5 keV).

A second method of quoting fluxes is based on the total energy received over a certain x-ray bandwidth. Units are commonly $erg\, cm^{-2} s^{-1}$, or $keV\, cm^{-2} s^{-1}$. Energy fluxes are directly proportional to photon fluxes only if the sources being compared have the same spectra.

Spectral density is usually expressed in units of $keV\, cm^{-2} s^{-1} keV^{-1}$, although the SI unit $W\, m^{-2} Hz^{-1}$ is convenient for comparisons with other spectral ranges. Energy spectra are commonly plotted on logarithmic axes, as spectral density in $keV\, cm^{-2} s^{-1} keV^{-1}$ against energy in keV. Another system in common use is to plot log photons $cm^{-2} s^{-1} keV^{-1}$ against log keV. Such photon number spectra appear steeper

16

than the corresponding energy spectra; they will not be used in this book in order to avoid confusion. An energy spectrum, $J = f(E)$, is equivalent to a photon number $N = 1/E \cdot f(E)$.)

To compare x-ray fluxes and spectra with data at other wavelengths, it is necessary to understand the other units used. Radio astronomers use units of $W\,m^{-2}\,Hz^{-1}$; frequently $10^{-26}\,W\,m^{-2}\,Hz^{-1}$ is called 1 jansky (Jy), or 1 flux unit. Visible-wavelength astronomers deal in terms of stellar magnitude. A zero-magnitude star yields a spectral density at the Earth of $3 \cdot 8 \times 10^{-23}\,W\,m^{-2}\,Hz^{-1}$ in yellow light (5500 Å wavelength). Fainter stars have larger magnitudes according the the logarithmic relationship,

$$m_s = -2 \cdot 5 \times \log_{10} \frac{f_s}{f_0},$$

where m_s is the stellar magnitude, f_s is the stellar flux, and f_0 is the zero-magnitude flux.

Finally, it is worth considering the significance of broad-band spectra plotted as $\log J$ ($W\,m^{-2}\,Hz^{-1}$) against $\log \nu$ (Hz). A spectrum of the form $J \propto \nu^{-1}$ yields equal energy fluxes per octave in any part of the spectrum, because there is more frequency bandwidth available at the higher frequencies. $10^{-29}\,W\,m^{-2}\,Hz^{-1}$ represents a weak radio source, but a strong x-ray source.

1.8.3. X-ray Luminosity

The luminosity of an astronomical source is a measure of the energy it radiates. Unlike the observed flux, it is a fundamental property of the source. Luminosity, L, is related to energy flux, F, according to the inverse square law

$$F = L/4\pi D^2,$$

where D is the distance of the source. In x-ray astronomy, as in most branches of astrophysics, luminosities are expressed in units of $erg\,s^{-1}$ (1 $erg\,s^{-1} = 10^{-7}$ W). The luminosity of the Sun = 1 $L_\odot = 3 \cdot 8 \times 10^{33}\,erg\,s^{-1}$.

Distances in astronomy are normally quoted in parsecs (pc), where $1\text{ pc} = 3\cdot1 \times 10^{18}\text{ cm} = 3\cdot1 \times 10^{16}\text{ m}$; $1\text{ kpc} = 10^3\text{ pc}$; $1\text{ Mpc} = 10^6\text{ pc}$. The nearest visible star lies at $D = 1\cdot3\text{ pc}$, and a distant galaxy may lie at 100 Mpc. Given F in $\text{erg cm}^{-2}\text{ s}^{-1}$, one must convert D into cm in order to calculate L in erg s^{-1}.

2. Observational Methods

2.1. Introduction

All observations in x-ray astronomy are made from platforms above the Earth's atmosphere.

In the energy range 0·2–20 keV, observations must be made from heights greater than about 120 km because of the strong absorption by the atmospheric gases. This involves the use of either a sounding rocket, whose trajectory takes it above 120 km for some four or five minutes, or of an Earth satellite, which can maintain a 500 km high orbit for some years. Higher satellite orbits are not favoured for x-ray astronomy because of the Earth's radiation belts; the charged atomic particles cause an increased background signal in x-ray detectors. Because the radiation belts reach lower at high geomagnetic latitudes, equatorial orbits are preferred for x-ray astronomy satellites, and sounding rockets are normally launched from sites well away from the poles of the Earth. The majority of the work reported in this book has been done in the energy range 0·2–20 keV from rocket and satellite platforms. Proportional-counter detectors are usually used.

In the energy range 30–150 keV, x-ray observations have been made from balloon platforms at heights of about 35 km. Work from these lower altitudes is possible because high-energy x-rays are less strongly absorbed by the atmosphere. Balloon flights have durations of hours, as against the minutes possible in a rocket flight. Satellites are, however, also used for work in the 30–150 keV range. In either case, scintillation counters are normally employed as detectors in this energy range.

2.2. Outline of Instrumentation for 0·2–20 keV

By far the most usual type of instrument used to date comprises one or more proportional-counter detectors placed behind a collimator. The resulting assembly is sensitive to x-rays coming from a direction defined by the collimator axis. This 'x-ray telescope' is scanned across the part of the sky in which the astronomer is interested. As an x-ray source passes through the field of view of the telescope, the detector registers an increased response. The response, after processing by on-board electronic circuits, is then transmitted to the ground by radio telemetry, together with information about the pointing direction of the spacecraft.

The pointing of the telescope is usually determined by the motion of the spacecraft. The motion may consist of the

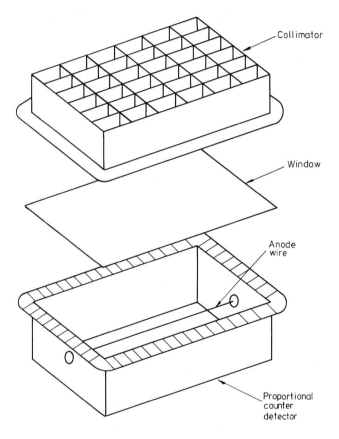

Figure 2.1. Exploded view of a typical astronomical x-ray detecting system.

vehicle rolling and precessing as in the early rocket surveys, as a simple roll about a more or less stationary axis as in the case of the Uhuru satellite, or as a programmed motion to cause the telescope to scan a region of special interest in the sky in the more sophisticated space vehicles. The pointing of the telescope is monitored using sensors mounted on the spacecraft, which detect directional features such as the Earth's horizon, the Earth's magnetic field, the Sun, the Moon and the visible stars. Gyroscopes are also sometimes used. It is normally a task for a computer to reconstruct the pointing history of the telescope from the attitude sensor signals which are transmitted to Earth by radio telemetry.

This basic technique is crude by the standards of the optical astronomer. Positional resolution is normally measured in minutes of arc of angle, some hundreds of times poorer than is obtained with a modest optical telescope. Moreover, the x-ray astronomer has, up till now, had to scan each astronomical object in turn, rather than being able to 'photograph' a region of sky. On the credit side, the x-ray astronomer does obtain some spectral information about every object he scans. A special technique has been evolved to improve on positional resolution using a modulation collimator. For x-ray objects close to the ecliptic, very precise positional information has been obtained by timing the occurrence of occultations by the Moon. Future x-ray telescopes are likely to employ focusing optics and imaging detectors, making them more like their optical counterparts. The first major satellite to carry such a system was HEAO-B, launched in November 1978. Such focusing telescopes, with high angular resolution and imaging capabilities, are likely to lead to major advances in x-ray astronomy.

2.3. The Proportional-counter Detector

Proportional-counter detectors are used in the majority of x-ray telescopes operating in the energy range 0·2–25 keV. The detector consists of an electrically conducting vessel, one side of which—the window—is made of a thin sheet of light metal, or plastic. It contains a gas mixture with about 90% partial pressure of a noble gas, such as xenon, argon or neon, and about 10% partial pressure of a polyatomic quench gas, such as methane or carbon dioxide; the total gas pressure being approximately atmospheric. An anode wire with a diameter of some 100–200 μm is supported between insulators at each end of the detector vessel. The anode wire is maintained at a potential of about +2000 V, and is connected to a low-noise electronic amplifier. Several such detectors can be combined to give a total sensitive window area of about

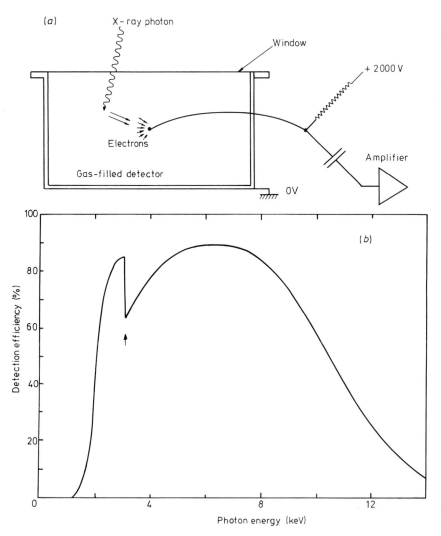

Figure 2.2. (a) Proportional-counter detector operation. (b) Photon detection efficiency of a typical argon-filled detector. The feature at about 3 keV is the K-edge of argon.

1000 cm^2. The detector vessel must be sealed, and must be strong enough to hold the gas against the vacuum environment of space.

The operation of the detector can be understood in the following way. An x-ray photon enters the detector through the window. It is photoelectrically absorbed in the noble gas. The resulting photoelectron has an energy comparable with that of the incident x-ray photon, and it loses its energy by ionising further atoms. The resulting free electrons are attracted towards the anode wire. As they approach this wire, they experience a strong electric field and are accelerated to energies at which they are capable of ionising further gas atoms. There is therefore a multiplication (or gas amplification) of the electrons which reach the anode wire. As the electrons reach the wire, they deposit a charge on it which is detected as an electrical pulse by the amplifier. The positive ions, meanwhile, drift more slowly towards the case (cathode) of the detector, and this charge motion also contributes to the signal on the anode. The function of the quench gas is to prevent further ionisation occurring in the gas when the positively charged ions reach the cathode. The electrical signal from a proportional-counter detector is normally shaped into a pulse of about 0.5×10^{-6} s duration by the amplifier. Because the response of a proportional-counter detector consists of discrete pulses, one for each x-ray detected, this response is often called the 'counting rate'.

The magnitude of the output pulse is proportional to the energy of the primary photoelectron, or photoelectrons, produced by the incident x-ray, since the number of ion pairs produced by the primary photoelectrons is proportional to the energy of the photoelectrons (about 30 eV being used to produce each ion pair). The gas amplification process is linear, the gain normally lying in the range 10^4–10^6; it is a function of detector geometry, gas pressure and composition, and anode voltage. There is some scatter in the pulse amplitude produced by primary photoelectrons of the same energy. This is a result of statistical fluctuations in the number of ion pairs produced. It follows that, whenever the energy of the primary photoelectrons equals that of the incident x-ray, then the pulse output is proportional to the x-ray energy, but with some scatter. The resolution of a typical proportional-counter amounts to about 16% in energy at 6 keV. This percentage energy resolution worsens at lower energies, varying as $(\text{energy})^{-1/2}$. Unfortunately, one further complication enters into the relationship between pulse height output and incident x-ray energy. When the x-ray ionises the first noble gas atom, it is likely to knock out the innermost electron from this atom. This photoelectron will possess the energy of the incident x-ray minus the energy with which the electron was bound into the atom. The atom will promptly adjust its electron shell, emitting one or more secondary x-ray photons in the process. If these photons are reabsorbed in the gas, further electrons will result, and the total photoelectron energy will nearly equal that of the incident x-ray. If, on the other hand, a secondary photon escapes from the counter, its energy will not be registered. The consequence of this is that a proportional-counter produces a double-peaked pulse height response to a mono-energetic incident beam of x-rays. The main peak contains pulses whose height is proportional to the energy of the incident x-rays. The 'escape peak' is less prominent than the main peak, and contains pulses whose height is proportional to the energy of the incident x-ray minus that of the escape photon. The energy of the escape photon amounts to about 3 keV when the gas is argon.

The quantum efficiency of a proportional-counter detector can reach values of 80% or more. It is limited on the low-energy side by the absorption of x-rays in the window, and on the high-energy side by the transparency of the gas to x-rays. Therefore, very low-energy x-rays cannot penetrate the window, and high-energy x-rays can pass straight through the gas without being absorbed. Commonly used window materials include plastic films of a few micrometres thickness which have some transmission even at 0·2 keV, and beryllium metal sheets of 50–100 μm thickness which transmit down to about 2 keV. Only the metal windows are sufficiently gas-tight for sealed detectors which must survive for more than a year in satellite telescopes. Commonly used gas mixtures are argon/CO_2 and xenon/CO_2, with a depth of about 50 mm at atmospheric pressure.

The proportional-counter detector responds not only to x-radiation, but also to charged particles and gamma-rays which pass through the detector. Above the Earth's atmosphere, the detector is immersed in the cosmic-ray particle and gamma-ray flux, so an unwanted source of background counts is always present. The following measures are usually taken to reduce this unwanted background to a minimum.

(1) Pulses with heights corresponding to x-ray energies above and below the useful range of the detector are electronically rejected before being transmitted to the ground by radio telemetry.

(2) Guard detectors are employed. These are additional detectors placed around the x-ray detector. A penetrating charged particle will then cause a response both from the x-ray detector and from one or more of the guard detectors, whereas an x-ray event will not affect the guard detectors. An anti-coincidence circuit ensures that any count seen by the x-ray detector is ignored whenever a simultaneous count is recorded by a guard detector.

(3) Risetime discrimination is employed. An electronic circuit distinguishes the risetime of events in the x-ray proportional detector. Background counts from gamma-rays and cosmic rays have longer risetimes than those from x-rays, and these long-risetime pulses are rejected by the circuitry.

All of these systems are commonly used in present-day x-ray astronomy packages, giving a background count rate from cosmic radiation of about $0 \cdot 01$ counts $cm^{-2} s^{-1}$. The response of the system to the isotopic astronomical x-ray background is minimised by restricting the field of view of the detector.

In summary, the proportional-counter detector has a high efficiency for the detection of x-rays, a high time resolution, a moderate energy resolution, and its background rate can be reduced to a reasonable level. It can be fabricated with sensitive areas of several hundred square centimetres; can readily be made sensitive over the energy range 2–20 keV; and by the use of ultra-thin windows the low-energy limit can be extended to $0 \cdot 2$ keV. For these reasons it has been the mainstay of x-ray astronomy instrumentation to date.

2.4. Collimators

A collimator consists of arrays of parallel metal plates placed directly in front of the detector window. The collimator also frequently supports the window against the gas pressure of the detector, when the detector is in its vacuum space environment.

The collimator will fully transmit to the detector x-rays incident along its axis, and will fully block radiation incident in a direction well away from the collimator axis. The response of the collimator is triangular in the following sense. Suppose that a parallel beam of x-rays (as from an x-ray star) is incident upon the collimator. As the collimator is rotated (see about its x axis figure 2.3(a)), the transmission of the beam of radiation will vary in a triangular fashion. Similarly, if the collimator is rotated about the other axis (y in figure 2.3(a)), the transmission pattern shown in figure 2.3(c) will result. The width of the collimator response is specified by

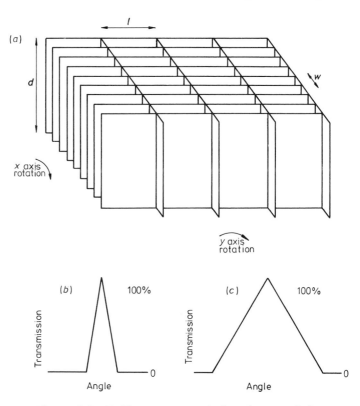

Figure 2.3. Collimator transmission characteristics.

the angular full width at half maximum (FWHM) of each triangle. Thus, using the dimensions d, l and w shown in figure 2.3(a), the FWHM about the x axis will be

$$\tan^{-1} w/d \simeq w/d \text{ rad,}$$

and the FWHM about the y axis will be

$$\tan^{-1} l/d \simeq l/d \text{ rad.}$$

The transmission is nonzero over a total angle of $2 \times$ FWHM. Practical considerations render it difficult to arrange for the FWHM of a collimator to be much smaller than $0 \cdot 25°$.

2.5. Determination of the Strength and Position of an X-ray Source

The rotation of the spacecraft carrying the x-ray telescope will cause the axis of the collimator to scan a path across the sky. For a spinning spacecraft, in which the collimator axis is normal to the spin axis, this path will be a great circle on the celestial sphere. Owing to the angular width of the collimator transmission pattern, a finite solid angle of sky will be viewed at any instant. A single x-ray star lying on the path of the collimator axis will yield a triangular counting rate response which has a width characterised by the appropriate FWHM of the collimator, and a height proportional to the strength of the star. A similar x-ray star of the same strength, lying above or below the path of the collimator axis, will yield a response of the same width, but of smaller height, if it lies within the field of view of the collimator. The collimator axis pointing direction, θ, at the peak of the triangle will therefore indicate the position of an x-ray star in one dimension only. If the x-ray source is not star-like, but is an extended nebula-like object, this will show up as a broadened response only if the width of the nebula approaches or exceeds the FWHM of the collimator. (The interpretation of many x-ray sources as star-like rests not on the analysis of the triangular response of the type of collimator under discussion; rather, it is deduced from the intensity variations of the sources.)

In order to locate an x-ray source in two dimensions with the type of instrument under discussion, it is necessary to observe the source at least twice. The ideal second scan is

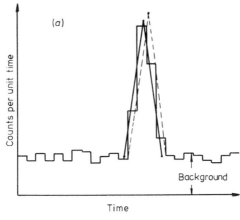

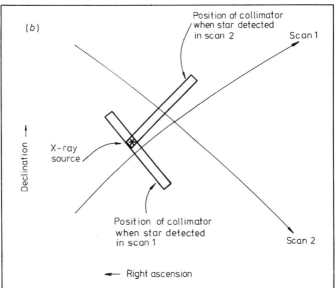

Figure 2.4. (*a*) Response of a collimated x-ray detector as a point source is scanned. (*b*) Location of an x-ray source from two intersecting scans.

made such that the path of the collimator axis intersects the first path at 90°. More elaborate arrangements of angled collimators can obviate the need for a second scan.

The analysis of an x-ray source observation as described above consists, therefore, of fitting a triangle to the observed data, and measuring its height and position. Both height and position will be determined with inherent errors which depend upon the strength of the source. The ideal triangular

28

response will be distorted by the random manner in which the x-ray photons and background events arrive at the detector. Suppose b is the average background counting rate per second, and s is the average counting rate from the source at the peak of the triangle. Then, if it takes a time, t, for the source to be scanned between half maximum points of the triangle, the total number of source counts accumulated will be

$$st \pm \sqrt{(s+b)t} = S \pm \sqrt{S+B},$$

where S and B are the average numbers of counts from source and background, respectively, in the time, t. $\sqrt{S+B}$ is the standard deviation resulting from the random arrival manner of the photons and background events. It follows, therefore, that the fractional error with which s can be measured will be of order $S/\sqrt{S+B}$, and will be better for stronger sources and smaller backgrounds. It can also be shown that the error in the positional location of the centre of the triangle will be of order $\text{FWHM} \times S/\sqrt{S+B}$, again showing the expected improvement for stronger sources and lower backgrounds. The source strength measured in this way will have to be corrected for the efficiency of the detector, and for the angle between the source and the path of the collimator axis.

2.6. Determination of the Spectrum of an X-ray Source

When an x-ray source is observed with a proportional-counter detector, the energy of each incident photon is translated by the detector into an electrical pulse height. It is therefore necessary to study the distribution of pulse heights (pulse height spectrum) in order to measure the x-ray spectrum of the source. In some systems, the pulses are analysed into height channels on board the spacecraft, and the spectrum (as an accumulated count in each channel) is transmitted to the ground every so often. In other systems, each x-ray pulse is individually transmitted to the ground with its height information retained, and then pulse height analysis is performed on the ground.

Once the pulse height spectrum from the source observation has been obtained, it is first of all necessary to remove the contamination due to background counts. This is done by subtracting the pulse height spectrum resulting from a corresponding exposure of the detector to a blank area of sky, from the pulse height spectrum of the source exposure.

Secondly, it is necessary to convert the resulting pulse height spectrum into the x-ray spectrum of the source. As a start, an x-ray energy is assigned to each pulse height channel. This is usually done with the aid of monochromatic calibration sources, such as ^{55}Fe (5·9 keV), carried on board the spacecraft, and exposed to the detector at intervals. In principle, it is then only necessary to divide each accumulated count by the corresponding detector efficiency at that energy to yield the spectrum in terms of received x-ray photons. In practice, this is an unsatisfactory procedure because it takes no account of the finite resolution and the escape peak characteristics of the detector. To overcome these difficulties the problem can be reversed. One starts by assuming a likely x-ray spectrum, and then calculates the pulse height spectrum with which the particular detector, with all its failings, would respond. The calculated pulse height spectrum is compared with that observed. The procedure is repeated for different assumed x-ray spectra until a satisfactory agreement between calculated and observed pulse height spectra is reached. The calculations involved are lengthy, but can easily be accomplished using a digital electronic computer. The method was first published by Gorenstein *et al* (1968) and has been widely used. The forms of x-ray spectra which are tried in this fitting procedure are usually continuum spectra of the bremsstrahlung plus emission lines, black-body and power-law types, together with a degree of low-energy cut-off. The reason for these choices is explained in Chapter 3.

The accuracy with which a source spectrum can be measured depends mainly on the signal-to-noise factor, $S/\sqrt{S+B}$, where S is the total number of counts accumulated from the source, and B is the number of background counts accumulated in the same time interval. The error in the determination of the spectrum shows up as an uncertainty in the temperature, or spectral index, and in the low-energy cut-off of the x-ray spectrum. Frequently, it is not possible to assign uniquely a spectral model (bremsstrahlung, black-body or power-law) when sources are only weakly observed.

The detection of spectral lines by x-ray astronomers is not well advanced, except in the Sun. Lines of iron at around 6·5 keV have been observed with proportional counters in several x-ray sources, including Sco X-1 and Cyg X-3. They appear as a small bump on the smooth pulse height spectrum, and have a width which reflects the limited resolution of the proportional-counter detector used. Searches for narrow lines have been made with Bragg crystal spectrometers, but with limited success to date. In the Bragg crystal spectrometer, the x-rays from the source are reflected by a crystalline material before they reach the detector. Only those x-rays whose wavelength, λ, satisfies the Bragg condition

$$2d \sin \beta = n\lambda$$

(where $n = 1, 2, 3$, etc, d is the crystal lattice spacing, and β is the Bragg angle) are reflected onto the detector, the remainder being absorbed by the crystal. Wavelength (or energy) resolutions of 1% are possible with these devices, but they are inefficient, and have been used with only the strongest celestial sources.

2.7. Other Types of Detector

The scintillation counter detector can be used to detect x-rays with energies greater than about 10 keV. It consists of a thin (roughly 6 mm thick) crystal of sodium iodide or caesium iodide doped with thallium, which is optically coupled to the face of a photomultiplier tube. When an x-ray photon enters the crystal, it is photoelectrically absorbed, as in the gas of a proportional counter. The photoelectron ionises further atoms in the crystal, and some of this ionisation energy is converted into a flash of visible light, or a scintillation, which is detected by the photomultiplier. The electrical pulse from the photomultiplier has a height which is proportional to the energy of the incident x-ray. So, like the proportional counter, the scintillation detector yields spectral information. A scintillation detector works at higher x-ray energies than a proportional counter, because it has a dense crystal made of a high-atomic-number material, which is more readily able to stop the penetrating x-radiation than the gas filling of a proportional counter. Scintillation detectors can be made sensitive to photons of energies even greater than 1 MeV (1 Mev = 1000 keV), if thick (100 mm) detector crystals are used; but such thick-crystal detectors have a poorer energy resolution around 30 keV than the thinner-crystal types. The scintillation detector also responds to cosmic ray particles, and is normally used with guard detectors in anti-coincidence to reduce the background count.

The channel multiplier is a device which finds a limited application in the detection of x-rays below about 3 keV. It is a vacuum device, and works in the space environment without needing a window. The x-rays fall onto a specially coated photo-emissive surface, which emits electrons. These electrons are then amplified by secondary emission as they are accelerated by an electrical field down the inside of a resistive tube. At the end of the tube, the electron signal is collected and fed to an amplifier. The channel multiplier possesses no energy resolution of the type found in proportional counters, and must be used in conjunction with filters placed in the x-ray beam in order to obtain spectral information. A channel matrix is a two-dimensional array of channel multipliers manufactured as one unit, and has some application in imaging x-ray systems.

The position-sensitive proportional-counter detector is a device which promises well for the future when imaging x-ray telescopes are flown in space. It consists of a normal proportional counter with either a resistive anode wire, or many parallel anode wires, or a combination of many resistive anode wires. With a resistive anode wire, the position of arrival of the x-ray along the wire may be found by comparing the size of the electrical signal from each end of the wire. With a multi-wire detector, the position of the arrival of the x-ray in a direction perpendicular to the wires may be found by comparing the signals from the various wires.

2.8. Modulation Collimators

A modulation collimator consists of two or more plane parallel grids of wires. The wires are usually spaced from each other by one wire's diameter, and are made of a high-atomic-number material such as tungsten, which is highly opaque to x-rays. The collimator functions as follows. Suppose that the front grid is illuminated by a parallel beam from an x-ray star, then the wires of the front grid will cast a series of shadows on the rear grid. If the shadows fall on the spaces between the wires of the rear grid, no x-radiation will reach the detector behind the rear grid. If, on the other hand, the shadows fall on the wires of the rear grid, the detector will receive half of the radiation incident on the front grid. Should the modulation collimator now be scanned across the source, the detector signal will be periodically modulated at a frequency which depends upon the scan rate, the pitch, p, of the wires, and the spacing, s, of the grids. The angle scanned between a maximum and an adjacent minimum of the modulation pattern amounts to $\frac{1}{2} \times \Delta \theta$, where

$$\tan \Delta \theta = p/s.$$

It is much easier to engineer a modulation collimator with a small angular resolution, $\Delta \theta$, than to make a conventional collimator with the same resolution.

A single scanning modulation collimator is of limited usefulness, because there is an ambiguity in the measured source position. Its only real use is to determine the size of a source.

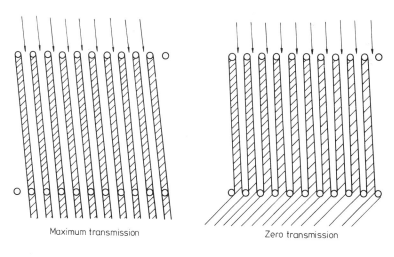

Maximum transmission   Zero transmission

Figure 2.5. The operation of a modulation collimator.

33

A star-like source will yield a 100% modulation depth; an extended source will give a smaller modulation depth, which depends upon the ratio of source extension to $\Delta\theta$. In this respect, a scanning modulation collimator behaves like an interferometer as used by radio astronomers.

Various adaptations of the modulation collimator make it a very useful device for the determination of source positions. The first adaptation, used to measure the position of Sco X-1 (Gursky *et al* 1966), employed a vernier technique. Two modulation collimator systems with slightly different values of $\Delta\theta$ were flown on the same rocket, and were scanned across the source. From the relative phase of the outputs, it was possible to eliminate most of the ambiguities in the source position. In order to measure the position in two dimensions, it was necessary to repeat the scan after rotating the instrument. In the variable-spacing modulation collimator (Adams *et al* 1972) the instrument is held steady with respect to the fixed stars, and the spacing between the grids is varied, such that the frequency of modulation depends upon the source position. This device is capable of locating several sources simultaneously; but it must be rotated, and the grid scan repeated, to obtain positional information in two dimensions. The rotation modulation collimator (RMC) is the most useful instrument of this type (Schnopper *et al* 1968, 1970). A single, fixed-spacing modulation collimator is axially pointed towards a fixed position in the sky, and is then rotated about its axis. Each point source in the field of view produces a rather complex modulation pattern; it is possible to analyse the data on a large computer to produce a two-dimensional map of the point sources in the region. This technique is well suited to spin-stabilised satellites, such as SAS-3. None of these techniques is well suited to the mapping of extended sources.

The most obvious advantage of the modulation collimator is its ability to locate x-ray sources with higher precision than is normally achieved with conventional collimators. This is often important in attempting to identify an x-ray source with an optical or infra-red counterpart. It has a sensitivity advantage over a conventional high-resolution collimator because when scanning a large region of sky with a modulation collimator, the source exposure is still large, even if working at very high resolution.

As well as the bright source Sco X-1, several of the stronger sources in the Galactic bulge have been accurately located using modulation collimators.

2.9. Focusing Systems

It is possible to focus x-rays using grazing-incidence reflection from polished metallic surfaces. The absorption properties of matter make lenses and normal-incidence mirrors useless at x-ray wavelengths. The basic x-ray optical component there-fore consists of a paraboloidal grazing-incidence reflector. Grazing-incidence reflection is more efficient at low x-ray energies, and in practice the devices are only useful at energies below 3 or 4 keV.

The grazing-incidence collector, utilising one reflection, is a crude device used to concentrate an incoming beam of x-radiation onto a small detector. Often the 'focusing' is done in only one dimension. The focusing properties are poor, but the use of a much smaller detector than is needed with a conventional collimator gives advantages in cosmic ray back-ground reduction.

The focusing x-ray telescope utilises two reflecting surfaces

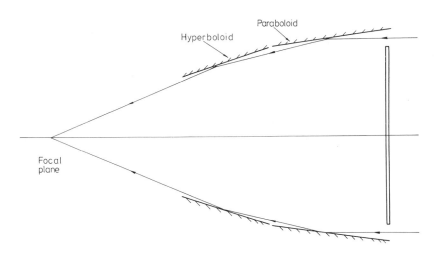

Figure 2.6. Cross section through a grazing-incidence reflecting tele-scope.

to provide high-quality images: one surface is paraboloidal and the other is hyperboloidal. Most of the observations obtained with such focusing telescopes to date have been of the Sun, since this yields a large enough flux to be recorded on photographic film. The combination of an imaging x-ray reflector with a sensitive-imaging proportional-counter detector promises a great future for observational x-ray astronomy. HEAO-B, launched in November 1978, is the first satellite to carry an instrument of this type, and its results are eagerly awaited. The details of the optics of a focusing x-ray telescope are described in Giacconi *et al* (1969).

2.10. Well Known X-ray Satellites

A small number of x-ray satellites have provided the majority of the observations reported in this monograph. Uhuru, the first United States x-ray observatory, was launched late in 1970, and carried simple collimated proportional counters which were scanned by the spin of the spacecraft. Changes in the orientation of the spin axis permitted the whole sky to be scanned. Ariel V was a similar satellite, instrumented by the British, which carried a number of instruments in addition to the Uhuru-like sky survey instrument. OSO-7 and ANS were rather similar. SAS-3 was instrumented with a rotating modulation collimator, and it has been used to provide accurate source positions for a number of x-ray objects. HEAO-A is the latest, and the largest, of the sky survey x-ray satellites.

Copernicus, also known as OAO-C, carried a small focusing collector, which has been used to make detailed studies of some known sources. HEAO-B (which was launched in late 1978) carries a full focusing x-ray telescope, and is likely to make a major impact on the study of x-ray sources. It will be capable of obtaining pictures of a quality which approaches that attainable with an optical telescope. Coupled with this imaging capability will be a sensitivity which far surpasses anything that has been available up to the present. Observations obtained with HEAO-B are likely to lead to a much better understanding of the faint extragalactic x-ray sources described in Chapter 6.

3. Theoretical Arguments in X-ray Astronomy

3.1. Introduction

In this chapter, arguments commonly used to interpret observations in x-ray astronomy are discussed. The accreting binary model of x-ray stars to be discussed in Chapter 4 is now so widely accepted that a description of the model is included in this chapter.

3.2. Source Variability

A common feature of Galactic x-ray sources, other than supernova remnants, is that their x-ray brightness varies on timescales of hours, minutes, and, in some cases, seconds. Cygnus X-1 has been seen to vary on a timescale of milliseconds. Even the extragalactic source in the radio galaxy Centaurus A has varied appreciably in intensity in the space of one week.

These intensity variations can be used to place limits on the size of the x-ray emitting regions as follows. Suppose that the source region has a simple spherical form with a radius R, such that the x-ray luminosity per unit volume is constant. Now suppose that for some reason the central 1% of the emitting volume decreases drastically in luminosity. Unless the majority of the x-ray emitting volume promptly follows this downward trend in luminosity, no appreciable change in brightness will be seen by the observing instrument. The luminosity change must therefore by communicated to the rest of the source region in some way. Since no signal can travel faster than c (the velocity of light), it will take a time of at least $t = R/c$ for the outer layers of the emitting region to follow the trend of the centre. The observed signal, which is the sum of contributions from the whole volume, will then decrease slowly over a time of about t. Therefore, if an x-ray source is seen to fluctuate appreciably in intensity (that is, by 20% or more) on a timescale of t, it follows that the radius of the emitting region cannot exceed about $R = c.t$. This conclusion is equally valid for variable radio, or visible light, sources.

X-ray sources which show fluctuations on a timescale of one second must therefore have radii smaller than $R = c \times 1 = 3 \times 10^{10} \times 1 \text{ cm} = 3 \times 10^{5} \text{ km}$. To put this number in perspective, the radius of our Sun is about $7 \times 10^{5} \text{ km}$. A variability on a timescale of one hour implies a source region of less than 10^{14} cm radius, and most Galactic sources show variability on this timescale. A nearby x-ray source may lie at a distance of say $500 \text{ pc} = 1 \cdot 5 \times 10^{21} \text{ cm}$. Hence, the angular diameter of a source varying on a timescale of one hour or less cannot exceed $2 \times 10^{14}/(1 \cdot 5 \times 10^{21})$ rad, or much less than 1 arc second. It is on the basis of their variability that many of the Galactic x-ray sources are classified as star-like.

3.3. X-ray Emission Mechanisms and Source Spectra

Various mechanisms have been proposed to account for the intense x-ray emission of astronomical objects. The main suggestions are: (i) thermal radiation from bodies at temperatures of some millions of degrees kelvin; (ii) synchrotron radiation by energetic cosmic ray electrons moving under the influence of a magnetic field; and (iii) inverse Compton scattering of cosmic ray electrons colliding with starlight or microwave photons. Thermal radiation may take the form of either bremsstrahlung from an optically thin gas, or of blackbody radiation from an optically thick object.

It has been a common practice to interpret observed x-ray spectra in terms of one or other of these emission mechanisms. Observed spectra usually show the following features:

(1) a fairly sharp cut-off towards the lower energies, and

(2) a continuum which decreases with increasing x-ray energy, E, either according to a power law of E, or according to an exponential law of E.

The low-energy cut-off is usually interpreted in terms of absorption by cool material in the line of sight (see § 3.10). A

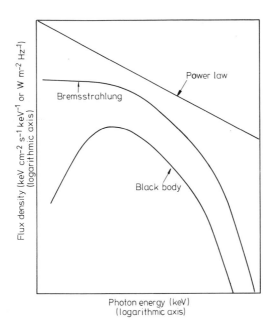

Figure 3.1. Standard assumed x-ray source spectra.

continuum which decreases with E according to a power law is frequently interpreted as having been generated by a synchrotron, or inverse Compton process, by electrons having a power-law energy spectrum. A continuum which falls off exponentially with E is usually interpreted as having been generated by a thermal process. Bremsstrahlung radiation is inferred to be present if the spectrum is flat at low energies, since a black-body spectrum shows a distinct peak. These interpretations are rough and ready, but generally seem to be on the right lines. The reader should, however, be aware of the following confusing cases.

(a) It is possible to construct a bremsstrahlung model when a range of temperatures is present, which can produce an apparent power-law spectrum over a limited energy range.

(b) It is possible to construct a synchrotron, or inverse Compton model, using a non-power-law cosmic ray electron energy spectrum, which can produce a roughly exponential x-ray spectrum.

(c) It is difficult to distinguish between, on the one hand, a black-body spectrum and, on the other, a bremsstrahlung spectrum which is appreciably cut off at low energies by the absorption of intervening cool matter, unless it is possible to distinguish the spectral emission lines which are usually produced in a bremsstrahlung source.

According to our present understanding of the nature of cosmic x-ray sources, the thermal production mechanisms are more common than the synchrotron and inverse Compton mechanisms. Accordingly, the thermal mechanisms—bremsstrahlung and black-body—will be dealt with first.

3.4. Thermal Bremsstrahlung (Free–Free) Radiation

Hot gases at temperatures above 10^7 K emit bremsstrahlung x-rays if they are optically thin. (Optically thin implies that the gas is insufficiently thick and dense to absorb appreciably its own radiation. This means that the spectrum of x-rays observed is the same as the spectrum during their production.) A hot gas emits principally by three processes—bremsstrahlung, bound–bound emission, and free bound emission. The second two processes involve the presence of atoms with at least some of their electrons remaining in bound orbits about the nucleus. Gaseous plasmas with a normal astrophysical abundance of the elements (i.e. made mostly of hydrogen and helium and with traces of heavier elements whose abundances diminish with increasing atomic weight) are almost completely ionised at temperatures above 10^7 K. Therefore the major emission process which needs to be considered is bremsstrahlung.

Bremsstrahlung (free–free radiation) is the process by which electrons radiate as they pass through the attractive Coulomb field of the positive ions in the plasma. The spectrum and intensity of the radiation emitted by each unit volume of the gas is

$$B_\nu = 6\cdot2 \times 10^{-39} g \exp\left(\frac{-h\nu}{kT}\right) \frac{1}{\sqrt{T}} n_e^2,$$

where B_ν is the intensity in units of $\text{erg cm}^{-3} \text{s}^{-1} \text{Hz}^{-1} \text{sr}^{-1}$; g is the Gaunt factor, a slowly varying function of ν and with a value of order unity; h is Planck's constant in CGS units; k is

Boltzmann's constant in CGS units; T is the temperature in K; and n_e is the electron density in units of cm^{-3}.

If the emitting region has a volume, V, the total energy radiated can be written as

$$L_\nu = 4\pi \times 6\cdot2\times 10^{-39}\mathrm{g}\ \exp\left(\frac{-h\nu}{kT}\right)\frac{1}{\sqrt{T}}\int n_e^2\, \mathrm{d}V\ \mathrm{ergs\,s^{-1}\,Hz^{-1}},$$

where $\int n_e^2\, \mathrm{d}V$ is known as the emission measure. Integrating over the emitted spectrum, we can obtain an expression for the total luminosity

$$L = 1\cdot64\times 10^{-27}\ \mathrm{g}\sqrt{T}\int n_e^2\, \mathrm{d}V\ \mathrm{ergs^{-1}}.$$

Thermal bremsstrahlung therefore yields a spectrum which falls exponentially for photon energies, $h\nu$, greater than kT, but which remains almost constant for $h\nu$ much smaller than kT. A black-body spectrum behaves in a similar way at high energies, but cuts off at energies below kT following a ν^2 law. It is useful for the x-ray astronomer to remember that $h\nu = kT$ corresponds to $T \simeq 1\cdot2\times10^7\,K$, if $h\nu = 1\,keV$.

Two possible factors can cause a bremsstrahlung spectrum to cut off towards low energies. The first is absorption by cool matter in the line of sight to the x-ray source. This effect is very noticeable in sources lying at distances away of 5 kpc or more in the Galactic plane, and will be discussed in more detail in § 3.10. The second is the onset of free–free absorption in the emitting plasma itself. The photon frequency at which this occurs depends on the electron density and the volume of the emitting plasma; it usually lies well below the x-ray part of the spectrum. The observation of the onset of free–free absorption in the infrared part of the spectrum of Sco X-1 was used to deduce the electron density in this source.

The assumption that bremsstrahlung is the only emission mechanism of a hot gas is adequate to account for the (poorly determined) spectra of several Galactic x-ray sources, with resulting temperatures in the range 20–$200\times10^6\,K$. There are, however, two additional factors which should be taken into account when constructing hot gas models of x-ray sources. Firstly, any realistic hot mass of gas is likely to have

41

temperature gradients within it, so that the resulting x-ray spectrum would be the resultant of bremsstrahlung spectra characterised by different temperatures. Secondly, it is not quite true to say that, at x-ray temperatures, a cosmic plasma is completely ionised. Even at 10^8 K, iron retains one or two bound electrons. Consequently, the so-called bremsstrahlung spectrum should have superimposed on it spectral lines of Fe XXV and Fe XXVI. These lines, although broadened, have been observed in sources including Sco X-1 and Cyg X-3. At 10^7 K, several other elements are incompletely ionised, whilst at 10^6 K, the spectrum of a hot gas of cosmic composition is dominated by line radiation. (The solar corona is a well studied example, see Culhane 1977.) It follows that the older supernova remnants, such as Vela X and the Cygnus Loop, believed to be optically thin thermal x-ray emitters at temperatures of a few million degrees, are expected to have spectra which deviate markedly from the simple exponential law. Detailed calculations of the spectra expected from such plasmas have been published by Tucker and Koren (1971).

The basic physics of bremsstrahlung radiation can be illustrated in the following way (see, for example, Rose 1973). A hot gas contains electrons which have become detached from their parent atoms. Consider one such electron moving at a (non-relativistic) velocity, v. Suppose that it approaches a singly charged nucleus with an impact parameter, b, as shown in figure 3.2. The electrostatic attraction between the electron and the positive ion will cause the electron to move around the ion in a parabolic orbit. The electron will

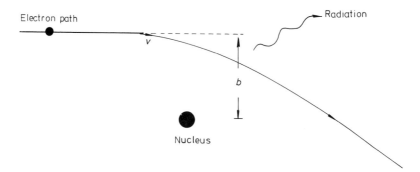

Figure 3.2. Mechanism of bremsstrahlung radiation.

experience an acceleration of roughly

$$a = (K_1 e^2)/b^2,$$

where e is the electron charge and K_1 is a constant, for a time of approximate duration

$$t = b/v.$$

Because the ion is much more massive than the electron, its acceleration may be neglected. The electron can therefore be thought of as radiating a pulse of radiation of duration, t. Fourier analysis of such a pulse shows that it has a prominent component at a photon frequency of approximately

$$\nu = \frac{1}{2\pi t} = \frac{v}{2\pi b}.$$

According to classical electromagnetic theory, the rate at which an accelerated electron radiates electromagnetic radiation energy is proportional to the square of its acceleration, that is,

$$dE/dt = K_2 a^2.$$

Substituting for a, and calling the constant term, K_3, we have

$$dE/dt = K_3/b^4.$$

The amount of energy radiated by one electron in the collision will therefore amount to

$$E = \frac{dE}{dt} \cdot t = \frac{K_3}{b^4} \cdot \frac{b}{v} = \frac{K_3}{b^3 v}.$$

The overall radiation of the plasma will be the sum of the effects of many electrons which have a range of velocities, v, and a range of impact parameters, b. Proceeding step by step, consider firstly the effect of the differing impact parameters that is, suppose that all the electrons have the same velocity, v. The probability, $P\,db$, that one electron will collide with a proton with an impact parameter in the range b to $b+db$ in one second depends upon the area of the impact annulus, $2\pi b\,db$, the proton density (which will be the same as the electron density, n_e), and the electron velocity, v, so that

$$P\,db = 2\pi b\,db\,n_e v.$$

43

The number of collisions per second, $N\,db$, with impact parameters in the range b to $b+db$ will also depend upon the electron density, n_e, so

$$N\,db = 2\pi b\,db\,n_e^2 v.$$

The amount of energy radiated per unit volume per second in the frequency range ν to $\nu+d\nu$ by these collisions will therefore amount to

$$I_\nu\,d\nu = EN\,db$$

$$= \frac{K_3}{b^3 v}\,2\pi b\,db\,n_e^2 v$$

$$= \frac{K_3 2\pi n_e^2\,db}{b^2}.$$

The frequency range, $d\nu$, can be related to the impact parameter range, db, using the relationship established above

$$\nu = \frac{v}{2\pi b}.$$

Differentiating,

$$|d\nu| = \frac{v\,db}{2\pi b^2}.$$

Substituting for db in the expression for I_ν

$$I_\nu\,d\nu = \frac{K_4 n_e^2\,d\nu}{v},$$

where K_4 collects together all the constant terms.

This equation states that the spectrum, I_ν, emitted by a plasma containing electrons with a single unique velocity would be flat. Clearly this spectrum cannot extend to infinitely high photon frequencies; the spectrum must cut off when the photon energy amounts to the total kinetic energy of the electron, that is, when

$$h\nu = \tfrac{1}{2}m_0 v^2,$$

where h is Planck's constant, and m_0 is the electron mass. Note that even in plasmas at x-ray temperatures, the

44

electrons are moving at non-relativistic velocities.

To complete the calculation, it is now only necessary to consider the effect of the velocity distribution of the electrons in a plasma, supposing them to be in thermal equilibrium at a temperature, T.

According to the Maxwell–Boltzmann distributions of velocity, the fraction of electrons which have a velocity with components in the range

$$v_x \quad \text{to} \quad v_x + dv_x$$
$$v_y \quad \text{to} \quad v_y + dv_y$$
$$v_z \quad \text{to} \quad v_z + dv_z$$

is given by

$$f(v)\, d^3v = \left(\frac{m_0}{2\pi kT}\right)^{3/2} \exp\left(\frac{-m_0 v^2}{2kT}\right) d^3v,$$

where

$$v = \sqrt{v_x^2 + v_y^2 + v_z^2} \quad \text{and} \quad d^3v = dv_x\, dv_y\, dv_z.$$

Electrons with velocities in the range $v\, d^3v$ will contribute

$$I_\nu\, d\nu = f(v)\, d^3v \frac{K_4 n_e^2}{v}\, d\nu \quad \text{for} \quad \nu < \frac{m_0 v^2}{2h}$$

to the observed spectrum. The total observed spectrum is obtained by integrating over the velocity distribution of the electrons.

$$I_\nu\, d\nu = K_4 n_e^2 \int_{\sqrt{2h\nu/m_0}}^{\infty} \frac{f(v)}{v}\, d^3v\, d\nu$$

$$= K_4 n_e^2\, d\nu \left(\frac{m_0}{2\pi kT}\right)^{3/2} \int_{\sqrt{2h\nu/m_0}}^{\infty} \frac{1}{v} \exp\left(\frac{-m_0 v^2}{2kT}\right) d^3v.$$

Rewriting the integral in spherical polar coordinates

$$d^3v = v^2\, dv\, d\theta \sin\theta\, d\Phi.$$

So

$$\int = \int_0^{2\pi} d\Phi \int_0^\pi \sin\theta \int_{\sqrt{2h\nu/m_0}}^{\infty} v \exp-\left(\frac{m_0 v^2}{2kT}\right) dv$$

$$= \text{constant} \times T \times \exp-\left(\frac{h\nu}{kT}\right).$$

45

So the spectrum expected from bremsstrahlung takes the form

$$I_\nu \, d\nu = K_5 n_e^2 \frac{1}{T^{1/2}} \exp - \left(\frac{h\nu}{kT} \right).$$

A more detailed treatment of bremsstrahlung yields

$$I_\nu \, d\nu = K_5 g(\nu) n_e^2 \frac{1}{T^{1/2}} \exp - \left(\frac{h\nu}{kT} \right),$$

where $g(\nu)$ is a quantum mechanical factor of order unity which varies only slowly with ν.

3.5. Black-body Radiation

A hot body which is optically thick will produce a black-body spectrum, whatever the underlying emission mechanism. This is because in an optically thick body the spectrum is influenced by absorption as well as by emission. Therefore, a bremsstrahlung source which is thick enough to absorb its own radiation will be observed to have a black-body spectrum. A hot neutron star will also have a black-body spectrum. The spectral form of black-body radiation depends only on the temperature, T, of the emitter. Planck's law gives the full spectral details:

$$\pi B_\nu = \frac{2\pi h\nu^3}{c^2 \{ \exp (h\nu/kT) - 1 \}},$$

where πB_ν is the energy per unit area per unit time per unit frequency bandwidth; h is Planck's constant; ν is the emitted photon frequency; c is the velocity of light; and k is Boltzmann's constant.

Planck's formula is cumbersome, but the following relations, which are easier to apply, can be derived from it. The photon frequency at which the peak emission occurs is given by

$$\nu \simeq 10^{11} T \text{ (Hz)}.$$

The total amount of energy emitted per unit area per second is given by

$$\pi B = \sigma T^4,$$

where σ is Stefan's constant.

46

Other useful data are given in Allen (1963). The derivation of the laws of black-body radiation is dealt with in many textbooks on physics.

As an example, we will calculate the basic parameters relating to a hot body which emits a black-body spectrum peaking in the x-ray region at 4 keV. An x-ray photon of energy 4 keV has a frequency $\nu \simeq 10^{18}$ Hz, so that $T \simeq 10^{18}/10^{11} = 10^7$ K, and the energy emitted per unit area amounts to $B = 5 \cdot 7 \times 10^{-5} \times 10^{28} = 5 \cdot 7 \times 10^{23}$ $erg\, cm^{-2}\, s^{\times 1}$. A bright Galactic x-ray object may have a luminosity of about 10^{37} ergs^{-1}. If it were a black-body emitter, its surface area would therefore need to be about $10^{37}/(5 \cdot 7 \times 10^{23}) = 1 \cdot 8 \times 10^{13}$ cm^2. If it were spherical in form, it would therefore need to have a radius of about $1 \cdot 2 \times 10^6$ cm $= 12$ km. It can be seen therefore that a very small body indeed could produce the x-ray luminosity of a typical x-ray star by black-body radiation. The interesting problem is to explain how such a body can be heated to such an enormously high temperature.

3.6. Production of High Temperatures—Accretion in Binary Systems

One type of model has successfully explained the production of the high temperatures needed to account for the majority of x-ray stars. The problem is not only to explain the production of temperatures of 10^7–10^8 K, but also to explain how the energy input can be maintained to support an x-radiation loss of 10^{36}–10^{38} ergs^{-1}. The model postulates the existence of a condensed object, such as a neutron star, in close proximity to a normal star. A neutron star has a mass roughly equal to that of the Sun, but is only about 10 km in radius. It is therefore made of extremely dense material, and is surrounded by an abnormally strong gravitational field. To produce very high temperatures, gaseous material must leave the normal star and fall into the gravitational potential well of the compact object. The gaseous material, infalling in towards the compact star, will be accelerated to a high velocity, and will tend to form into an 'accretion disc' around the compact star. The accretion disc is formed because the infalling material has angular momentum about the compact

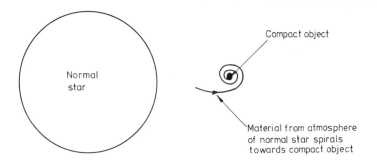

Figure 3.3. Accretion onto a compact object in a binary system.

star. The inner part of the accretion disc so formed will be at a temperature of tens or hundreds of millions of degrees kelvin. Material from the accretion disc which loses angular momentum and spirals onto the surface of the compact object will heat this up to a similar temperature. Both black-body and thermal bremsstrahlung radiation can be produced by such a system.

A simple calculation shows that x-ray temperatures can be produced by accretion onto a neutron star. Consider a proton of mass $m_p = 1\cdot67 \times 10^{-24}$ g falling from infinity onto the surface of the neutron star. The kinetic energy it will acquire is given by $E = GMm_p/R$, where M is the mass of the neutron star $(\simeq 2 \times 10^{33}$ g), R is the radius of the neutron star $(\simeq 10^6$ cm), and G is the universal gravitational constant $(= 6\cdot67 \times 10^{-8}$ CGS units). This gives a value for E of about 10^{-4} erg, if $T = 10^8$ K. A proton in a thermal plasma at a temperature T is characterised by a kinetic energy of $E = 1\cdot5\,kT$, where k is Boltzmann's constant $= 1\cdot38 \times 10^{-16}$ erg K^{-1}, giving $E = 2 \times 10^{-8}$ erg. The infalling protons therefore have more than enough kinetic energy to produce x-radiation.

Next, it is necessary to investigate the mass accretion rate which is required. The luminosity which must be explained in x-radiation amounts to about 5×10^{37} erg s^{-1}. The gravitational energy inputs amounts to

$$\frac{G\,M\,dM/dt}{R},$$

where dM/dt is the mass transfer rate, and M, R and G have the same values as used above. Equating the gravitational energy input to the x-ray luminosity, we find that $dM/dt \simeq 5 \times 10^{17}$ g s$^{-1} \simeq 10^{-8}$ solar masses per year. Such a mass transfer rate from the normal star to the neutron star is thought to be very plausible. If the normal star is a giant, or supergiant, it may lose mass through its stellar wind—larger mass loss rates can be directly observed by optical spectroscopy in many supergiants. Alternatively, mass transfer may occur through Roche lobe overflow; in this, the normal star expands as it evolves, until its outer layers come under the gravitational influence of the compact star. It is very unlikely that an isolated compact star could accrete sufficient material from the interstellar medium to become a steady x-ray source. In an accreting system, it is the size of the emitting region which determines its temperature, and therefore its peak emission wavelength. If the region has dimensions comparable with the size of a neutron star, then the peak emission will occur in the x-ray region of the spectrum.

An accreting x-ray source is a self-limiting system. If the luminosity were to rise above a certain value, then the radiation pressure of the x-rays would blow away any further accreting material, so diminishing the luminosity. There is therefore a maximum possible luminosity which is known as the Eddington limit. This is about 10^{38} erg s^{-1} for a compact object of 1 solar mass, and is proportional to the compact object mass. As the masses of neutron stars are believed to lie in the range 0·3–2 solar masses, it appears that many x-ray stars must be operating close to the Eddington limit.

There are three commonly discussed classes of compact object: white dwarf stars, neutron stars and black holes. All represent advanced stages of stellar evolution. A white dwarf is much larger than a neutron star, but has a comparable mass. It would disintegrate if it spun with a period shorter than a few seconds. A black hole is smaller than a neutron star and is more massive. It is an object so tightly compressed and with such a strong gravitational field that electromagnetic radiation is unable to escape from it. Black holes cannot therefore be seen to spin, and any observable effects such as the motion of a companion star or the presence of an accretion disc must be caused by the gravitational field at a distance. Neutron stars are generally favoured as the compact objects in x-ray binary systems, because they can easily explain the luminosity and the pulsations which are commonly observed. A compact object with a mass of more than twice that of the Sun cannot be a neutron star, but must be a black hole.

3.7. Production of X-rays by Synchrotron Radiation

Synchrotron radiation is generated when relativistic electrons are made to move in curved paths, and so are accelerated, by a magnetic field. A simplified explanation of the process, suggested by Ginzburg and Syrovatskij (1963), can be given as follows.

A relativistic electron moves with a velocity, v, which is close to the velocity of light, c. Its kinetic energy, E, will be equal to $m_0c^2(\gamma - 1)$, where γ is the Lorentz factor of special relativity $[\gamma = (1 - v^2/c^2)^{-1/2}]$, and m_0 is the rest mass of the electron. For highly relativistic electrons, γ will be much greater than unity, so that the kinetic energy, E, will be proportional to γ. The inertia of the electron will also increase because of the relativistic motion, so the electron will behave as if it had a mass of γm_0. Consider the simple case in which the electron is moving in a magnetic field of strength, B, at right angles to the field direction. The electron will

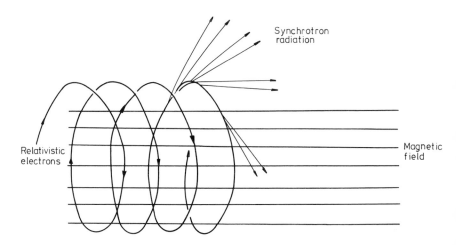

Figure 3.4. Mechanism of synchrotron radiation.

experience a force equal to evB, where e is the electrical charge on the electron. This force will cause the electron to move in circles about the lines of magnetic field. We may write down an equation relating the magnetic force on the electron to the product of its mass and centrifugal acceleration:

$$evB = \gamma m_0 v^2 / R,$$

where R is the orbital radius. The period of the electron in its orbit will be

$$T = 2\pi R / v = 2\pi \gamma m_0 / (eB).$$

Because the electron is being accelerated, it will emit electromagnetic radiation. If it were not moving relativistically, this radiation would be a sine wave at a frequency $\nu = 1/T$, and it would be called cyclotron radiation. But the relativistic motion of the electron introduces two important effects. Firstly, the electron only emits in a forwards cone with an opening angle of $\alpha = 1/\gamma$ rad. This means that an observer only sees a pulse of radiation while the electron occupies a fraction $\alpha/(2\pi)$ of its orbit. Secondly, when the observer sees this pulse of radiation, the electron is moving straight towards him, so that the duration of the observed pulse is Doppler-shifted by an amount $1 - v/c$. The observer therefore sees a pulse whose duration is approximately

$$t = T\frac{\alpha}{2\pi}(1 - v/c).$$

Substituting in terms of γ and B,

$$t = \frac{2\pi \gamma m_0}{eB} \cdot \frac{1}{2\pi\gamma} \cdot (1 - v/c),$$

and because $v/c \approx 1$,

$$(1 - v/c) \cdot (1 + v/c) = (1 - v^2/c^2) \approx 2 \cdot (1 - v/c),$$

so

$$1 - v/c \approx \tfrac{1}{2} \cdot (1 - v^2/c^2) = \tfrac{1}{2} \cdot \gamma^{-2};$$

thus

$$t \approx \frac{m_0}{eB2\gamma^2}.$$

Now if this train of pulses is analysed into its frequency

components by Fourier analysis, the dominant frequency component will occur at $\nu = 1/(2\pi t)$. Hence, the observed radiation will peak at a frequency $\nu \simeq \gamma^2 eB/(\pi m_0)$. A more careful derivation yields $\nu = 0 \cdot 07 \, \gamma^2 eB/m_0$.

If the electron is not moving at right angles to the magnetic field—as will generally be the case—the above formula is still valid, provided that B is replaced by $B_\perp$, the component of the magnetic field at right angles to the electron motion. The electron now moves in a helical orbit about the magnetic field lines.

The rate at which an electron of energy $E = \gamma m_0 c^2$ loses energy to synchrotron radiation is given by $-dE/dt = bE^2 B^2$, where b is a constant which depends upon the units used. A full treatment in relativistic electrodynamics is required to derive this formula.

The synchrotron theory has been used to account for the radio emission of supernova remnants and radio galaxies, and has been adopted to explain the x-ray emission of the Crab Nebula. The strength of the magnetic field in the Crab is about $5 \times 10^{-4} \, \text{G} = 5 \times 10^{-8} \, \text{Wb m}^{-2}$. Electrons which produce $\nu = 10^{18} \, \text{Hz}$ (4 keV) photons must have a Lorentz factor given by

$$\nu = 0 \cdot 07 \gamma^2 eB/m_0.$$

Using $e = 1 \cdot 6 \times 10^{-19} \, \text{C}$ and $m_0 = 9 \cdot 1 \times 10^{-31} \, \text{kg}$, we find that $\gamma = 4 \times 10^7$ or alternatively, $E = 2 \times 10^{13} \, \text{eV}$. The production of x-rays by the synchrotron mechanism therefore requires electrons of extremely high energies. In the Crab Nebula, electrons with energies of 10^8–$10^9 \, \text{eV}$ produce radio emission.

Synchrotron radiation is normally recognised by two observational features: (1) the radiation is linearly polarised, and (2) the spectrum of the radiation follows a power law $I(\nu) = k\nu^{-\alpha}$. The polarisation of the radiation follows immediately because the electrons are only accelerated at right angles to the magnetic field, so that the electric vector of the polarised radiation lies in the direction of the electron acceleration. That the spectrum of the radiation is a power law results from the power-law nature of the energy spectrum of cosmic rays. This latter point may be illustrated as follows.

Suppose that the electron spectrum is such that the number

of electrons with energies between E and $E+dE$ is $N(E)\,dE$. Then its power-law nature can be expressed in the form $N(E) = N_0 E^{-m}$, where m is a constant. Electrons of energy E will have a Lorentz factor, γ, given by $E \simeq \gamma m_0 c^2$. An electron with energy E will lose energy to synchrotron radiation at a rate given by $-dE/dt = bE^2 B_\perp^2$, where b is a constant, and will give rise to photons of frequency ν, where $\nu = aB_\perp E^2$. Photons of a frequency in the range ν to $\nu+d\nu$ will be generated by electrons in the energy range E to $E+dE$, with an intensity given by

$$I(\nu)\,d\nu = (-dE/dt)\,.\,N(E)\,dE.$$

From now on, we shall use a series of undefined constants, K_1, K_2, etc, as the spectral shape is not dependent on the value of the constants. Substituting for dE/dt and for $N(E)$,

$$I(\nu)\,d\nu = K_1 E^2 B_\perp^2 E^{-m}\,dE.$$

It is now necessary to express dE in terms of $d\nu$. We have

$$\nu = aE^2 B_\perp$$

whence

$$d\nu = 2aB_\perp E\,dE.$$

Substituting for dE in the expression for $I(\nu)$

$$I(\nu)\,d\nu = K_2 EB_\perp E^{-m}\,d\nu,$$

and substituting for E in terms of ν

$$I(\nu)\,d\nu = K_3 B_\perp (\nu/B_\perp)^{(1-m)/2}\,d\nu$$
$$= K_3 B_\perp^{(m+1)/2} \nu^{(1-m)/2}\,d\nu.$$

Thus it can be seen that an electron energy spectrum with an E^{-m} dependence leads to a synchrotron spectrum with a different, but related, spectral index

$$I(\nu) \propto \nu^{-\alpha},$$

where

$$\alpha = \frac{m-1}{2}.$$

3.8. Production of X-rays by the Inverse Compton Process

Inverse Compton radiation is generated when a relativistic electron collides with a low-energy photon. The electron

Figure 3.5. Inverse Compton scattering, illustrating the case of maximum energy transfer.

loses kinetic energy, which is transferred to the photon. If the low-energy photon has a frequency ν_1 and the resulting photon has a frequency ν_2, then these two quantities are related by $\nu_2 \simeq \gamma^2 \nu_1$. This may be illustrated as follows.

In order to avoid mathematical complexity, the special case in which the photon is scattered through 180° will be treated. Firstly, consider normal Compton scattering, in which an x-ray photon is scattered by an electron initially at rest. Suppose that, before the scattering, the photon has a frequency, ν_1', and a wavelength, $\lambda_1' = c/\nu_1'$, and that, after scattering, it has a wavelength, λ_2'. The energy of the photon of wavelength λ' is hc/λ', and its momentum is h/λ' in the direction of travel. Prior to the collision the electron is at rest, and so possesses zero kinetic energy and momentum. The electron will recoil with a velocity s' after the collision. Because the recoil velocity is likely to be relativistic, the expressions $(\Gamma' - 1)m_0 c^2$ and $\Gamma' m_0 s'$ must be used for the kinetic energy and momentum of the electron (where $\Gamma' = (1 - s'^2/c^2)^{-1/2}$). We may now write down the equations which express the conservation of energy and momentum in the collision.

Energy $\quad hc/\lambda_1' = hc/\lambda_2' + (\Gamma' - 1)m_0 c^2$

Momentum $\quad h/\lambda_1' = \Gamma' m_0 s' - h/\lambda_2'.$

Solving these equations, remembering that Γ' is related to s', we find

$$\lambda_2' - \lambda_1' = 2h/(m_0 c).$$

Inverse Compton scattering is the same interaction as above, but viewed in an inertial frame in which the electron is moving relativistically. The wavelengths of the two photons

54

involved will therefore appear Doppler-shifted by an amount which depends upon v, the initial velocity of the electron. Using the fully relativistic Doppler shift relation, the incident photon will appear to be redder in the observer's rest frame

$$\lambda_1 = \lambda_1' \sqrt{\frac{1 + v/c}{1 - v/c}}$$

and the scattered photon will appear to be blue-shifted

$$\lambda_2 = \lambda_2' \sqrt{\frac{1 - v/c}{1 + v/c}}.$$

But, from the above analysis of Compton scattering

$$\lambda_2' - \lambda_1' = 2h/m_0 c,$$

so that

$$\lambda_2 \sqrt{\frac{1 + v/c}{1 - v/c}} - \lambda_1 \sqrt{\frac{1 - v/c}{1 + v/c}} = \frac{2h}{m_0 c}.$$

For a highly relativistic electron

$$\gamma = (1 - v^2/c^2)^{-1/2},$$

and $v \simeq c$, so that $1 + v/c \simeq 2$. It follows that

$$\sqrt{\frac{1 - v/c}{1 + v/c}} = \sqrt{\frac{1 - v^2/c^2}{(1 + v/c)^2}} \simeq \sqrt{\frac{1 - v^2/c^2}{4}} = \frac{1}{2\gamma},$$

and therefore

$$\lambda_2 = \frac{\lambda_1}{4\gamma^2} + \frac{h}{\gamma m_0 c}.$$

Now for all cases of astrophysical interest

$$\frac{h}{\gamma m_0 c} \ll \frac{\lambda_1}{4\gamma^2}.$$

Hence

$$\lambda_2 \simeq \frac{\lambda_1}{4\gamma^2},$$

or, using $\nu_1 = c/\lambda_1$ and $\nu_2 = c/\lambda_2$,

$$\nu_2 \simeq 4\gamma^2 \nu_1.$$

If one considers collisions at different angles, one obtains the

55

mean result

$$\nu_2 \simeq \gamma^2 \nu_1.$$

The rate at which electron energy is converted into high-energy photon energy is proportional to the energy density, ω, of the low-energy photons

$$-dE/dt \propto \gamma^2 \omega.$$

Two sources of low-energy photons are usually considered when discussing the production of x-rays. Starlight photons with $\nu = 6 \times 10^{14}$ Hz interacting with $\gamma = 35$ ($1 \cdot 5 \times 10^7$ eV) electrons will produce 4 keV (10^{18} Hz) x-rays; and so will microwave background photons (5×10^{10} Hz) interacting with $\gamma = 5000$ ($E = 2 \cdot 5 \times 10^9$ eV) electrons.

The x-ray spectrum expected from the inverse Compton process will be a power law whose index is related to the spectral index of the electron energy spectrum.

3.9. The Origin of Cosmic Ray Electrons

That relativistic cosmic ray electrons exist in many parts of the Universe is amply demonstrated by radio astronomy, but the mechanism by which these cosmic rays are produced is open to conjecture. Supernova explosions, pulsars, and activity in the nuclei of galaxies are commonly discussed in this context, but simply on the basis that these objects seem likely to have sufficient energy resources. In the case of the Crab Nebula, it is firmly believed that the pulsar NP 0532 continuously replenishes the supply of cosmic ray electrons.

3.10. Loe-energy Spectral Cut-off

Spectral cut-off at low x-ray energies is observed in many sources. It is caused by photoelectric absorption of the x-rays by material in the line of sight to the source. This material may include both the interstellar gas, which is distributed diffusely through our own and other galaxies, and material surrounding the x-ray source itself.

Photoelectric absorption of x-rays by material which is not ionised increases rapidly towards lower energies. For hydrogen gas, this increase follows a simple relationship with the

56

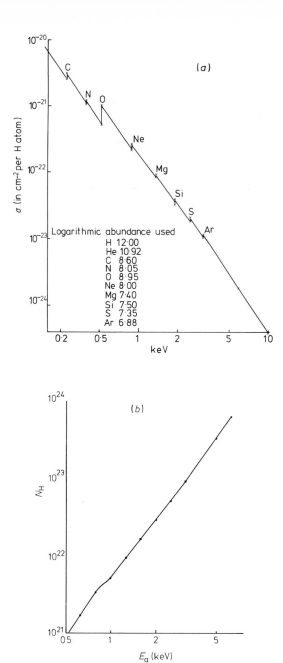

Figure 3.6. (*a*) The variation of the photoelectric absorption cross section of interstellar material with x-ray energy. (*b*) N_{H} against E_a curve. From Brown and Gould (1970).

energy. However, the interstellar medium consists, in terms of numbers of atoms, of roughly 89% hydrogen, 11% helium, and 0·1% heavier atoms such as carbon, nitrogen and oxygen. Because these heavier elements absorb x-rays much more readily, they must be taken into account when calculating the absorption properties of interstellar material. Figure 3.6(a) shows how the x-ray absorption cross section varies with energy for a gas of interstellar composition (according to Brown and Gould 1970). The absorption cross section, σ, is defined in such a way that a beam of radiation of intensity I_0 incident upon a region of material of column density N_H hydrogen atoms per unit area, will emerge from the column with an intensity I, where

$$I = I_0 \exp{-(\sigma N_H)}.$$

The discontinuities in the absorption coefficient are the K-edges of the heavier elements, and occur at the energies at which x-rays are just able to remove the innermost electron from an atom.

Figure 3.6(b) shows the relationship between the cut-off energy, E_a, and the depth of the column of absorbing material responsible. E_a is the energy at which the source spectrum is attenuated by a factor e (i.e. 2·712). E_a is easy to define for power-law and exponential spectra, but confusion can easily arise for black-body spectra, which themselves cut off towards low energies.

The low-energy cut-off in the spectrum of an x-ray source can be used to place an upper limit on the distance of a source which lies in the Milky Way. One assumes that the mean hydrogen density in the Galaxy is 0·5 atoms cm^{-3}. Only an upper limit to the distance is obtained because the source may also suffer from low-energy absorption. The non-uniformity of the interstellar medium makes the method rather unreliable. Extragalactic sources at high Galactic latitudes do not suffer appreciable x-ray absorption in the material of our own Galaxy, and any low-energy cut-offs observed are intrinsic to the extragalactic source.

4. Star-like X-ray Sources

4.1. Introduction

The majority of the brighter x-ray sources lie in the Galaxy and are star-like in nature. A few similar objects have been detected in the Magellanic clouds. They are known to be of only stellar dimensions by their variability on short time-scales. In addition to the random variability which is shown by almost all of the star-like sources, very significant periodic changes have been detected in the x-ray signals from some of them. Pulsations with periods of seconds or minutes, and a regular on–off modulation with a period of a few days, are commonly observed. These periodicities, and other evidence, clearly indicate that the x-ray sources are situated in close binary star systems, with one of the stars being a compact object.

Binary star systems have been known to optical astronomers for many years. In most such systems, the two stars in orbit around each other are too close together to be resolved at the telescope as separate stars. Their mutual gravitational attraction causes them to move in orbits about each other, in the same way that the Earth orbits the Sun. Binary stars are usually recognised by the periodic changes in the Doppler shift of their spectral lines caused by their orbital motion. In some binary star systems, one star periodically passes in front of and eclipses the other; such eclipsing binaries show periodic diminutions in brightness. Whether or not a spectroscopic binary exhibits eclipses depends on the size of the stars, their separation and the orientation of the plane of their orbit relative to the line of sight. Some x-ray stars show eclipsing binary behaviour in x-rays as the periodic on–off modulation, some show a spectroscopic binary type of behaviour in the spectra of their optical counterparts, and some show a spectroscopic binary type of Doppler shift in their x-ray pulsation periods.

A binary star model for x-ray sources, involving a normal star and a compact star in a close binary configuration, is capable of explaining the large observed luminosities of x-ray stars. These luminosities are typically between one thousand and one hundred thousand times the luminosity of the Sun, and are explained as the energy released when gas from the normal star falls into the intense gravitational field of the binary companion compact object. In fact, it is difficult to construct any plausible alternative model to account for the large energy output of x-ray stars.

In §§ 4.2 and 4.3, the better known x-ray stars which show strong evidence of their binary nature are described. Sections 4.4 and 4.5 deal with x-ray transients, Galactic bulge x-ray sources and 'bursters', all of which are probably binary in nature. Section 4.6 describes some low-luminosity x-ray sources, which are not necessarily binaries.

4.2. Eclipsing Binary X-ray Stars

4.2.1. Centaurus X-3

The pulsating behaviour of the x-ray source Centaurus X-3 makes it, together with Hercules X-1, one of the most significant discoveries of the satellite x-ray astronomy era. The discovery came about in early 1971, when a lot of interest was focused on the fast, but non-periodic, intensity variations in Cygnus X-1. The Uhuru team undertook a survey of the Milky Way in order to search for further examples of x-ray sources which varied in intensity on a timescale of seconds or so. Cen X-3 showed up as having such fluctuations in intensity, and it was soon recognised that these variations could be fitted to a curve with a 4·8 second period (Giacconi *et al* 1971). The radiation from Cen X-3 shows a high degree of pulsed modulation, dropping to about 30% of its peak intensity in the valleys between the pulses. There was evidence from these observations indicating that the pulse shape is complex, with a secondary peak following about 90° in phase behind the main pulse.

Further studies of the source have led to the conclusion that the x-ray emitting 'pulsar' resides in a binary star system (Schreier *et al* 1972). Radio pulsars exhibit two important

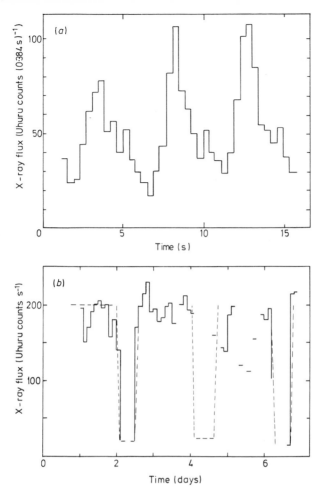

Figure 4.1. (a) The pulsations of Cen X-3. (b) The eclipses of Cen X-3. Data from Giacconi *et al* (1971) and Schreier *et al* (1972).

features: their pulsation frequency is virtually constant, but the slight change which can be detected indicates that the pulsations are slowing down. It was therefore natural to study the constancy of the frequency of the Cen X-3 pulsations, to see how the x-rays from Cen X-3 compared with the radio emission from a normal pulsar. It was found that the pulsations from Cen X-3 are quite unlike those from a normal pulsar; the frequency changes by several parts in 10 000 over the duration of a day or so. Cen X-3 shows other behaviour which would distinguish it from a normal pulsar. At times,

61

the x-ray intensity drops to about 10% of the normal level. The key discovery of the Uhuru observers was that these seemingly random dips in the x-ray intensity could be understood as being made up of two components—a periodic 2·087 day modulation, in which the source is in its high state for 1·5 days only, and an occasional 'extended low' behaviour, in which the source is not seen in its high state for some weeks. The object is never seen to be in a high state when the 2·087 day law predicts a low, but can be in a low state when the 2·087 day law predicts a high. The interpretation of the periodic behaviour is that the x-ray emitting object is gravitationally bound to another star, and is periodically eclipsed from view as it orbits behind that star. The extended low behaviour requires another explanation, such as the occasional presence of dense clouds of gas around the system to absorb the x-rays.

The eclipsing binary explanation was elegantly supported by the interpretation of the frequency changes of the pulsating x-rays in terms of Doppler shifts of the radiation from the orbiting x-ray pulsar. When the x-ray pulsar is approaching

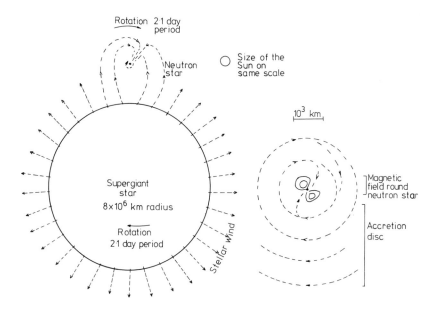

Figure 4.2. *Left*: The Cen X-3 binary star system. *Right*: A greatly magnified view of the region around the neutron star.

the observer, the pulsation frequency increases and when the pulsar is receding from the observer, the pulsation frequency decreases. Exactly this effect was observed by the Uhuru workers, and it was found to occur periodically with the same period as deduced from the eclipses. The eclipse occurs in the interval between the decreased frequency pulsations (x-ray source receding) and the increased frequency pulsations (x-ray source approaching). Paradoxically, the Doppler shift observations give a more accurate measure of the eclipsing period than the eclipse observations themselves. We therefore understand the x-ray pulsar itself in Cen X-3 to have a relatively constant frequency; the changes in frequency observed are related to its motion within the binary system. Long-term observations of the pulsations indicate that they are speeding up, in contrast with the pulsations of radio pulsars, which are slowing down.

The x-ray spectrum of Cen X-3 was measured by Cooke and Pounds (1971) before the binary nature of the source was discovered. The spectrum showed a moderately flat peak from 3 to 6 keV, and fell off steeply towards both high and low energies. Such a spectrum resembles that of a black body at about 15×10^6 K, but can also be fitted to a bremsstrahlung model with a low-energy cut-off parameter, E_a, equal to 3 or 4 keV. Such large values for E_a are not easy to explain on the basis of interstellar absorption, and would have to be attributed to absorption within the source. Baity et al (1974) have reported observations of Cen X-3 at energies of 7–30 keV, at which energies the eclipses are still apparent. They report that the flux falls rapidly with increasing energy, and that there is some evidence for variations in the spectral slope.

More detailed studies of the variation in the x-ray emission from Cen X-3 through its eclipse cycle have revealed two interesting features. Parkinson et al (1974) observed the source as the x-ray object came out of eclipse. Their instrument on board the satellite Copernicus comprised detectors sensitive to the energy ranges 0·7–2·0, 1·4–4·2 and 4–12 keV. The 1·4–4·2 keV signal came out of eclipse more than an hour after the high-energy signal. This indicates that the low-energy x-rays are absorbed in the atmosphere of the

63

normal star, whereas the high-energy x-rays can penetrate the atmosphere. This suggests that the normal star has a very extended atmosphere. Pounds *et al* (1975) observed dips in the x-ray intensity of Cen X-3 in the middle of the uneclipsed part of the 2·087 day cycle just after the source emerged from an extended low state. These dips are thought to be caused by dense gas moving into the line of sight; the gas is most probably material which has left the normal star and forms an 'accretion wake' on the other side of the compact star.

It should be emphasised that although the nature of Cen X-3 is best discussed from a study of its regular pulsations and eclipses, it also exhibits random variations in intensity, like almost all stellar x-ray sources. When the object is in its uneclipsed state, it often exhibits flares in which the x-ray intensity can double in a few hours.

The x-ray measurements which have been made on Cen X-3 give an excellent opportunity to make deductions about the nature of the astronomical object without appeal to optical or radio data. The 4·8 s pulsations indicate that the x-ray emitting region must be small compared with 4·8 light seconds = $1·5 \times 10^6$ km. The constancy of the pulsations requires that they are clocked by a rotating object. A neutron star is a compact object which can rotate once every five seconds without flying apart, and which has a strong enough gravitational field to generate x-rays by accretion. The Doppler shift observations can then be utilised to define the parameters of the binary system by an argument which runs as follows.

(i) The Doppler shift observations show an excellent fit to a sine curve. This means that the velocity of the condensed star is constant, and that the orbit about the centre of mass of the two stars is circular. The actual upper limit to the eccentricity of the orbit was determined as 0·05 by Schreier *et al* (1972). Other figures quoted in this section come from the same source.

(ii) The Doppler shift observations then indicate directly the projected radius of the condensed star about the centre of mass of the system. Thus the time delay of the pulses between phases 0·5 and 0·75 equals the light travel time over

64

the orbit radius r multiplied by sin i, where i is the inclination of the binary orbit axis to the line of sight. This was measured to be 39.75 light seconds $= 1.2 \times 10^7$ km, or about 16 times the solar radius.

(iii) The orbital velocity can be obtained by dividing the circumference of the orbit, $2\pi r$, by the orbital period, T,

$$v \sin i = 2\pi r \sin i / T = 415 \text{ km s}^{-1}.$$

Note that sin i still appears in the expression.

(iv) By applying Newton's laws of motion to the system, a relationship can be obtained involving the masses of the stars in the system. If m is the mass of the condensed star, and M is the mass of the non-x-ray star, then the gravitational force acting to bind the two stars together will be

$$GMm/(r + mr/M)^2,$$

where G is the gravitational constant, and $(r + mr/M)$ is the separation between the centres of the two stars. This force can be equated to the mass of either star multiplied by its inward acceleration; for the x-ray star this is mv^2/r. So

$$\frac{mv^2}{r} = \frac{GMm}{r^2(1 + m/M)^2}.$$

Rearranging

$$\frac{M^3}{(M+m)^2} = \frac{v^2 r}{G},$$

and using the orbital period

$$T = \frac{2\pi r}{v},$$

then

$$\frac{M^3}{(M+m)^2} = \frac{(2\pi)^2 r^3}{GT^2}.$$

But r has not been determined, only $r \sin i$, so it is only possible to evaluate

$$\frac{M^3 \sin^3 i}{(M+m)^2} = \frac{(2\pi)^2 r^3 \sin^3 i}{GT^2}.$$

Substituting for the known values of T and $r \sin i$, it is found that

$$\frac{M^3 \sin^3 i}{(M+m)^2} = 3 \cdot 07 \times 10^{34}\, g = 15 \text{ solar masses.}$$

(v) Sin i has appeared as an unknown in the above expressions for radius, velocity and mass ratio. Although sin i cannot be uniquely determined from the x-ray data, it is possible to place some limits on its value from the eclipse observations; i must be greater than zero, otherwise eclipse and Doppler shift phenomena would not be observed. The eclipse lasts for about a quarter of the orbital period, from which it can be argued that i lies between 45 and 90°. The following are therefore limits on the parameters of the binary system:

Orbital velocity of x-ray star, v
$$415 \text{ km s}^{-1} < v < 588 \text{ km s}^{-1}.$$

Orbital radius of x-ray star, r
$$1 \cdot 19 \times 10^7 \text{ km} < r < 1 \cdot 69 \times 10^7 \text{ km}.$$

Mass function
$$3 \cdot 07 \times 10^{34}\, g < \frac{M^3}{(M+m)^2} < 8 \cdot 81 \times 10^{34}\, g.$$

The optical identification of Cen X-3 was delayed until early 1974, when Krzeminski (1974) located a star which appeared to show optical intensity variations with the same period as the x-ray eclipses. This star lies within the positional error box determined by Parkinson *et al* (1974), and has a spectral type of BOIb and a visible magnitude of 13. The distance is estimated to be around 10 kpc. No pulsations can be detected in the visible. The blue supergiant star will have a mass in the range 16–22 solar masses; this can be determined by comparing its spectrum with spectra of stars of known mass. Applying restrictions to the formula above, the mass of the compact x-ray star is found to lie between 0·6 and 1·1 solar masses. This mass is fully consistent with the compact star being a neutron star.

Observations of Cen X-3 therefore indicate that it is a binary star system composed of a neutron star spinning on its axis once every 4·8 seconds, and a blue supergiant. Mass is

66

transferred from the supergiant to the neutron star by the stellar wind blowing off the supergiant, and the neutron star must be strongly magnetised in order that its rotation can modulate the x-ray flux. An accretion disc must form around the neutron star because the stellar wind has considerable angular momentum around the neutron star. The inner parts of the accretion disc will be hot enough to contribute to the x-ray emission. The outer parts of the accretion disc will emit visible radiation, but this will be too faint to be seen against the intense light from the supergiant star.

4.2.2. Hercules X-1

Her X-1 was the second pulsating x-ray star of the Cen X-3 type to be discovered. Tananbaum *et al* (1972) reported that the Uhuru satellite had observed pulsations with a period of 1·24 seconds, that the object showed an eclipsing binary behaviour with a period of 1·7 days, and that the object was not always detectable when the eclipsing binary law predicted that it should have been. Her X-1 lies at a high Galactic latitude; it is therefore favourably placed for optical observations as it is not seriously obscured by the dust in the Milky Way. The x-ray object was identified with the variable star HZ Her in 1972, and this star has been studied intensively since then.

The pulsation period of 1·24 seconds makes the frequency some four times greater than that of Cen X-3. It appears that most of the radiation is pulsed. Doxsey *et al* (1973) have obtained a detailed pulse profile of the object which shows it to have a double-peaked structure. The two peaks are separated by about 90° in phase angle, and there is evidence for some radiation at other phase angles. The pulsations are attributed to a mechanism clocked by the rotation of a magnetised neutron star once every 1·24 seconds. (A rotating white dwarf is ruled out because it would fly apart if it rotated at this speed.) Her X-1 has also been observed to pulsate very slightly in the visible, the pulsating component amounting to some 0·2% of the total visible radiation (Davidsen *et al* 1972, Middleditch and Nelson 1973). The visible pulsations are not always detectable, implying that the pulsed flux shows variability over and above that caused by eclipses. The pulsations in x-rays from Her X-1 show Doppler shifts in synchronism with the eclipses, and show a very gradual increase in frequency.

The eclipsing binary behaviour of Her X-1 shows many similarities to that of Cen X-3, except that the mass of the normal star appears to be much smaller. Tananbaum *et al* (1972) found that the duration of the eclipse was only 0·24 days, or 14% of the 1·7 day eclipse period. This suggests that the radius of the orbit of the small x-ray object about the companion star is larger compared with the companion star radius than is the case for Cen X-3. The Doppler shift observations are consistent with those expected from an eclipsing binary system with circular orbits. The following parameters can be derived for the system using the arguments which were described in the previous section on Cen X-3.

Projected orbital velocity of x-ray star	$v \sin i = 169 \cdot 2 \text{ km s}^{-1}$.
Projected radius of x-ray star orbit	$r \sin i = 3 \cdot 9 \times 10^6 \text{ km}$.
Mass function	$M^3 \sin^3 i/(M+m)^2$ $= 1 \cdot 69 \times 10^{33} \text{ g}$ $= 0 \cdot 85 \text{ solar masses.}$

The mass function is much smaller than in the case of Cen X-3, indicating that the normal star is one of low mass. The orbit of the x-ray star (neutron star) about the centre of mass of the two stars is also much smaller than in the case of Cen X-3, indicating that Her X-1 is a more compact system, with a smaller normal star.

The entry into and exit from eclipse occur more rapidly for Her X-1 than for Cen X-3; Her X-1 takes about 12 minutes, compared with 60 minutes for Cen X-3. Taking into account the smaller orbital velocity of Her X-1, this indicates that the companion star of Her X-1 has an atmosphere which is considerably less extended than the atmosphere of the supergiant in Cen X-3. This is as might be expected for a star of only a little more than the mass of the Sun, and with a radius comparable with that of the Sun.

Her X-1 exhibits pre-eclipse dips; the x-ray intensity drops and then rises again just before the x-ray eclipse occurs. The intensity does not always drop below the level of detectability in these dips, and the spectrum shows a larger than normal degree of low-energy absorption at these times. It is likely that these dips are caused by the presence of an absorbing gas stream in the binary system.

Long-term studies of the x-radiation from Her X-1 (see, for example, Holt *et al* 1976) reveal an extended low behaviour in which the source is only 'on' for about 11 days out of 35. Faint x-ray emission can, however, sometimes be detected when the source is 'off'. This extended low behaviour contrasts with that of Cen X-3 in two ways. Firstly, Her X-1 is 'off' for well over 50% of the time, and secondly, the 'off' states of Her X-1 occur regularly with a period of 35 days. The transition into the 'on' state occurs more rapidly than the transition into the 'off' state. The 'on' transitions should yield the best method for determining the precise extended low period, but Giacconi *et al* (1973) found that the 'on' transitions are not periodic, whereas the remainder of the profile of the 'on' state does repeat from cycle to cycle. It appears that the time at which a turn-on occurs depends on both the phase in the 35 day cycle and the phase in the 1·7 day eclipse cycle. The precise timing of the pre-eclipse dips in the eclipse cycle also depends on the phase of the 35 day cycle. The cause of the 35 day cycle is not clear. One possible explanation is that the neutron star spin axis precesses with a 35 day period such that the cone of x-radiation moves out of the line of sight to the Earth. It is difficult, however, to explain the fluctuations in the turn-on time in this model. Another possibility is that the accretion process takes place in 35 day bursts. In this case, the fluctuations in turn-on time can be plausibly explained in terms the geometry of the accreting gas columns.

The spectrum of Her X-1 is extremely hard; that is, it falls off rather slowly towards higher energies. The Uhuru observers (Giacconi *et al* 1973) obtained best fits with a power-law index of 0·0, or with a bremsstrahlung spectrum with a temperature greater than 150×10^6 K. Both fits required varying degrees of low-energy absorption depending on the phase of the 1·7 day and the 35 day cycles. Clark *et al* (1972) observed the source over a wider energy range of 1–60 keV, and obtained a bremsstrahlung temperature of 700×10^6 K, or a correspondingly hard power law. Ulmer *et al* (1972) confirmed the hardness of the spectrum in the 7–26 keV range. It is clear, therefore, that Her X-1 is at a higher x-ray temperature than Cen X-3.

Two emission lines have been observed in the spectrum of Her X-1. A weak iron line was detected at about 6·5 keV by Pravdo *et al* (1977). This line is consistent with there being a normal abundance of iron in the gases surrounding Her X-1. A more interesting spectral line in Her X-1 is the narrow 58 keV emission feature reported by Trumper *et al* (1978). This line is attributed to cyclotron radiation and is produced when non-relativistic electrons move in a strong magnetic field. The strength of the magnetic field needed to produce a cyclotron line at 58 keV is 5×10^{12} G. This is the strength of the magnetic field expected near the surface of the spinning neutron star.

The optical identification of Her X-1 was achieved after a great flurry of activity in 1972. Owing to the failure of the star sensors on Uhuru, the Uhuru workers were only able to

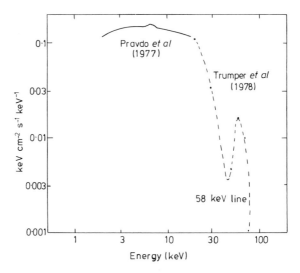

Figure 4.3. The x-ray spectrum of Her X-1. Note that the two observations were made at different times, and that the source is variable.

measure the position of the x-ray star approximately. Using the OSO-7 satellite, Clark *et al* (1972) located the source within an error circle of radius 0·3°. Liller searched this region of sky and found a 13th magnitude star with a large ultraviolet excess. Lamb and Sovari then observed the star and detected weak pulsations at the x-ray pulsation frequency. Bahcall and Bahcall observed photometric variations with the same period as the eclipses of the x-ray star. Finally, Liller obtained a precise photometric period in agreement with the x-ray eclipse period by studying old photographs of that region of sky. All this activity took place over a few months, and was communicated by means of IAU telegrams, a system designed to warn astronomers of the appearance of short-lived phenomena such as novae and comets. The outcome was that Her X-1 was positively identified with the variable star HZ Her. The x-ray position was subsequently improved by Doxsey *et al* (1973), who located the source within an error circle of radius 0·5 arc minutes using a rocket-borne modulation collimator. HZ Her lies comfortably within this error circle.

Optical photometry of HZ Her (Davidsen *et al* 1972, Bahcall and Bahcall 1972, Forman *et al* 1972) yields a light curve which shows a deep minimum in phase with the x-ray eclipse. The amplitude of the light curve is nearly two magnitudes through a B filter. The ultraviolet excess is very large ($U-B = -0·7$) except when the x-ray star is in eclipse, when it takes a more normal value ($U-B = 0·4$). This suggests that the ultraviolet radiation is coming from the vicinity of the x-ray star. The light variation itself is probably not caused so much by the eclipse of the x-ray star, as by the rotation of the normal star which is heated on one side by the strong x-ray flux.

Spectroscopic observations of the star HZ Her have been made by Crampton and Hutchings (1972), Bopp *et al* (1973) and Crampton (1974). The spectral type of the underlying star, presumably the normal star in the system, changes with the 1·7 day phase of the system. At minimum light, the star has a spectral type of about F5, whereas at maximum, the spectral type is that of a much hotter B-type star. This can be understood if the normal star has a spectral type of F5

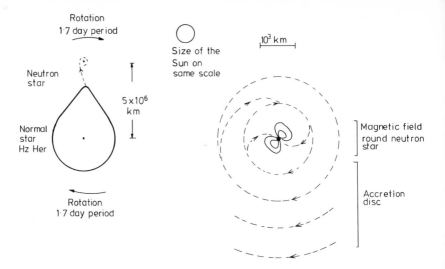

Figure 4.4. *Left*: The Her X-1 binary star system. *Right*: A greatly magnified view of the region around the neutron star.

appropriate to its mass, and the B-type spectrum is caused by x-ray heating. The spectroscopic Doppler shift curve is consistent with that to be expected from the motion of a distorted F-type star. Weak emission lines, variable in intensity, have been observed. The mass of the normal star is estimated to be about 2·5 solar masses, which makes the x-ray star mass about 1·3 solar masses. The normal star is much less massive than the corresponding star in Cen X-3. The x-ray star mass is quite consistent with its being a neutron star.

Jones *et al* (1973) have studied the photometric variations in HZ Her using optical plates dating back to 1890. They have established that the 1·7 day orbital period has remained constant over the last 80 years. The light curve, showing a deep minimum with an amplitude of 1·8 magnitudes, is not always seen; sometimes the light curve shows two minima per cycle with an amplitude of about 0·3 magnitudes. This change can be understood if one supposes that the distorted normal star is being seen in different aspects.

The distance of HZ Her has been estimated by Forman *et al* (1972). From arguments based on the details of binary star theory, they consider that the normal star in the system must be an F3 star with an absolute visual magnitude of +0·6. They then obtain a distance modulus of 13·8, corresponding to a distance of about 6 kpc, and an x-ray luminosity of about 10^{37} erg s^{-1}. This distance places the binary system well above the plane of the Galaxy.

Observations of Her X-1 demonstrate convincingly that it is a binary star system in which one star is a normal star slightly more massive then the Sun, and the other star is a spinning magnetised neutron star. The normal star is less massive and much less luminous than the supergiant in Cen X-3, and the two stars are closer together than those in Cen X-3. The normal star does not have a stellar wind which is strong enough to transfer material to the neutron star, and it is necessary to suppose that the normal star has expanded so that some of its outer atmosphere is gravitationally attracted towards the neutron star. This mechanism of mass transfer is known as 'Roche lobe overflow'. As is the case for other x-ray binaries, an accretion disc emitting both x-rays and visible light must form around the neutron star.

4.2.3. Other X-ray Eclipsing Binaries

A number of other stars which show an eclipsing binary behaviour in x-rays are known. Most appear to lie in the spiral arms of our Galaxy, with the exception of SMC X-1, which is in the Small Magellanic Cloud. Some show pulsations, and most have supergiant companions like Cen X-3. Vela X-1 (4U 0900 −40) shows pulsations with a much longer period than Cen X-3 or Her X-1 (282 seconds), and it is known as a 'slow rotator'.

4.3. Other Well Known X-ray Binary Stars

4.3.1. Scorpio X-1

The object Scorpio X-1 was observed on the rocket flight which discovered the first celestial x-ray sources (Giacconi *et al* 1962). It is the brightest object in the x-ray sky (apart from the Sun), and has therefore been studied intensively. Its apparent brightness results from its proximity to the Sun, rather than from anything unusual about its intrinsic x-ray luminosity as compared with other x-ray stars. The brightest star in the visible sky, Sirius, is also bright because it is close to us, and is similarly unremarkable in luminosity as visible stars go. It must be emphasised, however, that Sco X-1 is 100 times more distant than Sirius, and emits very much more energy in x-rays than Sirius emits in the visible.

The x-ray spectrum of Sco X-1 has been well studied. Between 2 and 10 keV, a good fit can be obtained by a bremsstrahlung spectrum of the form $I(E) = $ constant. $E^{-0.3} . \exp(-E/kT)$, where the characteristic temperature, T, varies between the limits 40×10^6 and 100×10^6 K (Gorenstein *et al* 1968). These spectral variations are real, so that the spectrum is characterised by one temperature at one time, and by another temperature at another time. The shape of the spectrum at energies below 2 keV also shows a marked variability (Grader *et al* 1970). This variable low-energy absorption suggests that the amount of absorbing material in the line of sight to the x-ray source is varying. The material probably comprises clouds of gas in motion around the x-ray source region. At energies above 15 keV, variability in intensity has also been observed (Lewin *et al* 1968), and the intensity generally lies above that which would be predicted by an extrapolation of the 2–10 keV bremsstrahlung spectrum. It would indeed be surprising if the whole spectrum of the object did fit a bremsstrahlung curve at a single temperature. The single-temperature fit would imply that the whole optically thin source region is isothermal. One further point worthy of note is the observation of Holt *et al* (1969) and Acton *et al* (1970) of a weak emission feature between 6·5 and 7 keV. This feature corresponds in energy with the expected emission from highly ionised iron at a temperature of $40 - 100 \times 10^6$ K.

The identification of Sco X-1 with its optical counterpart formed one of the early milestones in x-ray astronomy. The first problem was to assess the likely magnitude of the optical object, and the following argument was used. As the x-ray spectrum is strongly suggestive of a thin-body bremsstrahlung mechanism for generating the x-rays, then this mechanism should also be capable of generating longer wavelengths such as visible light. The approximate spectral expression $I(E) = $ constant $. \exp(-E/kT)$ could therefore be extrapolated to optical wavelengths, indicating that the object should give a visible magnitude of about 13. It should also have a large ultraviolet excess relative to the spectra of normal stars (whose intensities fall off in the ultraviolet). This argument gives a lower limit to the optical brightness—the presence of a companion star would make the system appear brighter in the visible. The density of stars of 13th magnitude or brighter in Scorpio is about 60 per square degree. It followed that, in order to have a chance of finding the object optically, it was necessary to measure its position with an accuracy of a few arc minutes or better. This was no mean objective in 1965, when the fan beam collimators used by rocket observers normally gave positional accuracies of a degree or so. Also, it was not known at the time that x-ray stars were binary in nature.

The position of Sco X-1 was measured in 1966 by the first modulation collimator to be flown on a rocket. The operation

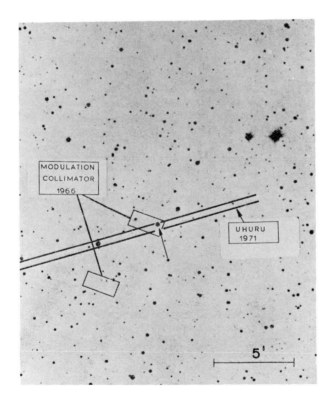

Figure 4.5. Early positional measurements of Sco X-1 shown superimposed on a sky photograph. The optical counterpart of Sco X-1 is arrowed.

of this device is described in Chapter 2. The measurement led to two significant results. Firstly, it showed that Sco X-1 had an angular extent of less than 20 arc seconds (Gursky *et al* 1966a), indicating that it was probably star-like. Secondly, the position was determined to lie within one of two error boxes, each of dimensions 1 arc minute by 2 arc minutes (Gursky *et al* 1966b). Sandage *et al* (1966) searched these regions of sky with an optical telescope for an object which was star-like, was 13th magnitude or brighter, and showed an ultraviolet excess. They found such an object, and suggested that it was the optical counterpart of Sco X-1. The identification has been amply confirmed by more recent positional measurements, and by the observation of correlations between the optical and x-ray variations in the object. The optical star is usually known by the x-ray name Sco X-1.

The distance of Sco X-1 can be estimated in a number of ways. Firstly, it lies at an unusually large angle from the Galactic plane as compared with other Galactic x-ray sources. From this angle of 24°, and if one assumes that it lies not more than 150 pc above the central plane of the Galaxy, then its distance must be less than about 150 cosec 24° pc = 400 pc. This argument is based on a dubious assumption about the distance above the plane, and would not work for Her X-1, for example. Sounder distance estimates have been made by studying the absorption lines present in the spectrum of the optical counterpart. The H and K lines of ionised calcium (Ca II) are frequently affected by the interstellar medium between a star and the Solar System. On the basis of radial velocity studies, Wallerstein (1967) confirmed that this is the case for Sco X-1. From a measurement of the equivalent widths of these lines, it is possible to compare the distance of Sco X-1 with that of a star whose distance has been measured by other means. The greater the amount of intervening matter, the greater the equivalent width of the interstellar line. Wallerstein (1967) showed that the equivalent width of the K line in the spectrum of the Sco X-1 optical counterpart is greater than that observed in the stars of the nearby Sco–Cen association. This places a lower limit of 270 pc on the distance to Sco X-1. Westphal *et al* (1968) estimated the distance to be 500–600 pc using a similar method. Various other optical astronomy methods have indicated distances between 170 and 1000 pc (Hiltner and Mook 1970). Studies of the low-energy x-ray spectrum have led to rather smaller distance estimates, but the method is rather unreliable, as explained in Chapter 3. The general consensus of opinion is that the optical distance measurements are the most reliable, and that the true distance of Sco X-1 lies in the range 200–1000 pc. If the distance is taken as being 500 pc, the x-ray luminosity of Sco X-1 amounts to 5×10^{35} erg s^{-1}.

Optical studies of Sco X-1 have been intensive. The intensity varies from 13th to 12th magnitude in an irregular manner. The magnitude may change by the full amount from night to night; it changes by 0·5 magnitudes in times as short as one hour; and a flickering of 0·02 magnitudes occurs on a timescale of minutes. Flares of amplitude 0·2 magnitudes and risetimes as short as 90 seconds are also seen to occur (Hiltner and Mook 1970). There is a correlation between the optical variations and the variable behaviour seen in x-rays—the source has two well defined states. In one state, the optical and x-ray intensities both remain relatively steady, and vary by no more than 20% over a one-day timescale. In the second state, both the optical and x-ray intensities show a more pronounced variability, and change by a factor of two in times as short as one hour. Both optically and in x-rays, the intensity is higher in this second highly variable state, and the x-ray spectrum is characterised by a higher temperature (White *et al* 1976). The two states, which each tend to persist for a few days, show a good correlation between x-rays and optical, but detailed, minute to minute correlations are not found. This explains the failure of the earlier rocket shots to measure any x-ray changes correlated with visible fluctuations. The observations suggest that the region producing the optical radiation is close to, but not coincident with, the x-ray source.

The only direct evidence for the binary nature of Sco X-1 comes from the visible-wavelength photometry of Gottlieb *et al* (1975) and spectroscopy by Cowley and Crampton (1975). The binary period is 0·787 days. No such periodic behaviour has been seen in x-rays.

Several attempts have been made to detect the spectral

77

lines in the x-ray spectrum of Sco X-1, which were suggested by the low-resolution proportional-counter spectrometer observations of Holt *et al* (1969). The motivation for these searches was to demonstrate beyond doubt that the emission mechanism is thermal. Tucker and Koren (1971) have calculated the detailed spectrum to be expected from a plasma at 50×10^6 K, and have shown that, for any reasonable assumption about the elemental abundances in the plasma, line features should be present in the spectrum. In particular, the lines of highly ionised iron, Fe XXV at 6·7 keV and Fe XXVI at 6·9 keV, should be observed. Using large-area Bragg crystal spectrometers, these lines have been searched for with sensitivities adequate to detect the most pessimistically predicted lines. These high-resolution searches which are only sensitive to narrow line features (Griffiths *et al* 1971, Stockman *et al* 1973) have yielded null results. On the other hand, Acton *et al* (1970) and Griffiths *et al* (1971) have confirmed the presence of a 6·7 keV feature at low resolution. The conclusion is that the line feature is present, but is severely broadened by electron scattering.

A variable radio source has been detected at the precise location of Sco X-1 (Ables 1969). This has a flux which varies between 0·2 flux units and 0·005 flux units (1 flux unit = 1 Jansky = 10^{-26} W m^{-2} Hz^{-1}). There are also two non-varying radio sources situated on either side of Sco X-1.

Theoretical models of Sco X-1 usually involve a normal star of low luminosity in a binary system, together with a compact object such as a neutron star. The accretion disc, which forms around the compact object, is responsible for much of the x-ray and visible radiation. Eclipses are not seen because of an unfavourable orbital inclination.

4.3.2. *Cygnus X-1*

Cygnus X-1 was one of the first x-ray sources to be detected. It is among the brightest of the x-ray objects in the sky in the 1–10 keV band, and at 100 keV it stands out as the strongest source. It was the first object in which x-ray variability was noticed. Such was its history until late in 1971, when it was identified with a seemingly normal BO supergiant star. Since

78

that date it has attracted more attention than any other x-ray star in the sky, because detailed interpretation of the observations leads to a model in which the compact object, around which the x-rays are produced, is a black hole. Cyg X-1 provides the best evidence yet for the existence of a black hole—a highly compact collapsed star with a surface gravity so strong that light cannot escape from it.

The variability of Cyg X-1 was the first feature to attract attention. The source was observed from the two rocket surveys conducted by the US Naval Research Laboratory group in 1964 and 1965 (Byram *et al* 1966). At the time of the second survey, the intensity in x-rays had dropped to a quarter of that observed in the first survey. More recently, the variability of the object has been studied in greater detail. Oda *et al* (1971) noted that the source varied on timescales of less than one second, and Schreier *et al* (1971) concluded that variations were present in their Uhuru data with timescales as short as 50 ms. Variability on a timescale of milliseconds was reported by Rothschild *et al* (1974), who flew a specially designed rocket instrument to detect it. Although there has been some argument about the validity of this last result, there is no doubt that Cyg X-1 is extremely variable on very short timescales. The only other x-ray source which shows such rapid random variability is Circinus X-1.

No pulsations of the type seen in Cen X-3 and Her X-1 have been detected in Cyg X-1. Neither does Cyg X-1 show x-ray eclipses, although there is a weak modulation with a period of 5·6 days (Sanford *et al* 1974, Holt *et al* 1976).

There are four points worthy of note in the x-ray spectrum of Cyg X-1. Firstly, the 2–10 keV portion of the spectrum shows a marked variability in slope. Secondly, there is a negligible degree of low-energy absorption present at energies as low as 1 keV. Thirdly, the source is the brightest in the sky at energies above 30 keV, and has a harder spectrum in this region than the Crab Nebula. Fourthly, the overall 1–100 keV spectrum is impossible to fit with either a single bremsstrahlung curve or with a single power law. The Uhuru observers (Tananbaum *et al* 1972) observed a dramatic change in the 2–10 keV spectrum. Fitting power-law models to these two states, they obtained power-law indices of 4·1

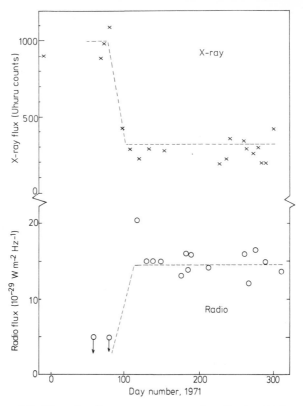

Figure 4.6. The simultaneous x-ray and radio level change in Cyg X-1. Data from Tananbaum *et al* (1972).

and 0·45, an enormous change by any standards. The intensity of the source was four times greater in the 2–6 keV band when the spectrum had the steeper slope, but the 10–20 keV intensity was greater by a factor of two when the 2–10 keV spectrum was flatter. The change from the one state to the other took place over a few weeks. The radiation seen below 10 keV has a spectrum which can be fitted with a bremsstrahlung form with a temperature of about 11×10^6 K when the source is in its brighter state, with the softer or steeper spectral form. A thermal spectrum does not fit the data when the source is in its less intense state with a harder spectrum. At higher energies, the spectrum is consistent with a power law with a spectral index of 1·0.

The study of the higher-energy x-radiation from Cyg X-1 has, in the main, been carried out by research groups different from those which study x-rays in the 0·25–30 keV band.

The gas-filled proportional counters used as detectors at around 5 keV become transparent to x-rays with energies above 20 or 30 keV and, in their place, it is necessary to use scintillation counter detectors in which the x-ray photons are stopped in a dense solid crystal of sodium iodide or caesium iodide. The photon fluxes at these higher energies are much smaller than in the 2–10 keV range, and consequently, the 300 second exposure of a sounding rocket flight is inadequate to make meaningful measurements. High-altitude balloons offer much longer exposure times of typically several hours, and attain an altitude which is great enough to observe the more penetrating x-rays. Accordingly, it was the research groups who had access to balloon vehicles who contributed the majority of the early observations of Cyg X-1 above 30 keV. Satellites are also now used to observe in this energy range. McCracken *et al* (1966) made the fist high-energy observation of Cyg X-1; since that date, more than 20 such balloon observations have been made (see review by Agrawal *et al* 1972). Many of the measurements extend up to 100 keV, and the majority of the spectra are consistent with a power law of spectral index of 0·9 or 1·0. Haymes *et al* (1968) observed the source up to an energy of 450 keV, and

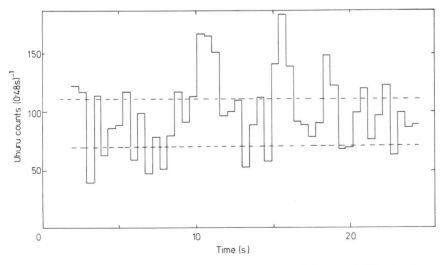

Figure 4.7. Rapid variations in the x-ray output of Cyg X-1. The broken lines indicate the maximum fluctuations to be expected from a steady source. Data from Schreier *et al* (1971).

81

claim that above 150 keV the spectrum falls off with a steeper slope. Cyg X-1 does show variability in intensity in the 30–100 keV range, but the spectral slope remains fairly constant.

Early attempts at the optical identification of Cyg X-1 were unsuccessful. Giacconi *et al* (1967b) searched the error box around the position of the source as determined by a rocket flight (Giacconi *et al* 1967a). They were looking for a star with a significant ultraviolet excess, following the method used to identify Sco X-1. They found no suitable candidate, although they were successful in finding the optical counterpart of Cyg X-2. It has since been shown that their position for Cyg X-1 was in error by about $\frac{1}{4}°$; this error may have been caused by the variability of the source.

The identification of Cyg X-1 in 1971 followed from another line suggested by studies of Sco X-1. Braes and Miley (1971) and Hjellming and Wade (1971) searched for and found a radio source coincident with Cyg X-1. The x-ray source position had by this time been improved to an accuracy of 0·5 arc minute by Rappaport *et al* (1971) using a rotating modulation collimator. The radio identification was made certain when it was noticed that a fourfold decrease in the x-ray flux (Tananbaum *et al* 1972) coincided within a week or two, with a large increase in the radio flux (Braes and Miley 1971, Hjellming 1973). The techniques of radio astronomy allowed much more accurate positional determinations than did the methods of x-ray astronomy. Accordingly, the radio source position was measured very precisely by Braes and Miley (1972) and Wade and Hjellming (1972), and it was found to coincide exactly with the position of a 9th magnitude star, HDE 226868. The identification of the x-ray source Cyg X-1 with a relatively bright star was therefore made by a most roundabout procedure.

The properties of the Cyg X-1 optical counterpart, HDE 226868, were first investigated by Murdin and Webster (1971), who found it to have the properties of a normal BOIb supergiant star at a distance of about 2 kpc. It does not show the ultraviolet excess which is shown by the optical counterparts of Her X-1 and Sco X-1; this is not too surprising, since the radiation would need to be of exceptional intensity to be noticeable against the strong emission from the supergiant. Doppler shifts in the spectral lines of HDE 226868 were interpreted by Webster and Murdin (1972) and Bolton (1972) to be those of a single-line spectroscopic binary system, in which only the lines of the brighter star in the binary system could be seen. The other star in the system must therefore be the compact object around which the x-rays are produced. Because total x-ray eclipses are not observed, the inclination, i, of the binary orbital axis to the line of sight must be significantly less than 90°. Using a method similar to that used to analyse the x-ray Doppler-shift data on Cen X-3, Webster and Murdin (1972) derived the following parameters for the system.

Projected orbital velocity of visible star
$$V \sin i = 64 \text{ km s}^{-1}.$$

Projected orbital radius of visible star
$$R \sin i = 4 \cdot 6 \times 10^6 \text{ km}.$$

Mass function
$$\frac{m^3 \sin^3 i}{(m + M)^2}$$
$$= 0 \cdot 12 \text{ solar masses}.$$

(M = mass of visible star; m = mass of compact star)

More refined work has since been reported, but the essential conclusion remains the same. If the supergiant star has a normal mass for its type of about 23 solar masses, then the compact star around which the x-rays are produced must have a mass considerably in excess of the two solar mass limit for a neutron star or a white dwarf. The compact star must therefore be a black hole. This conclusion has been confirmed by Bolton (1972), Brucato and Kristian (1973) and Hutchings et al (1973). HE II emission lines have been observed in the system, which vary out of phase with the spectral lines of the B star (Hutchings et al 1973), and have been attributed to emission from streams of gas within the binary system. Light variations of 5·6 days have been observed by Lester et al (1973), and have been interpreted as tidal distortions of the supergiant star by the compact star. These observations again do not change the conclusion about the mass of the x-ray star, which must lie in the range 10–20 solar masses.

Cyg X-1 must therefore be a binary system rather like Cen X-3, in which a blue supergiant star has a compact companion. Mass transfer from the supergiant will take place owing to the stellar wind from the supergiant. The difference is that in Cyg X-1 the compact object is a black hole, rather than a neutron star as found in most of the x-ray binaries. The x-radiation must come entirely from the accretion disc surrounding the black hole.

The black hole model for Cyg X-1 is so important that it is wise to question the assumptions which have led to the model. The probability of a chance coincidence being responsible for the identification with HDE 226868 is negligible. The optical properties of the B star could possibly mimic those of a much less massive star, as was suggested by Trimble *et al* (1973). If this were the case, then Cyg X-1 would be less luminous, and would have to be as close as 1 kpc. Both Bregman *et al* (1973) and Margon *et al* (1973) have studied the optical data from that region of the sky, and have concluded that a distance as small as 1 kpc is completely ruled out. Perhaps the most dubious assumption made is that the Doppler shift data can be interpreted in terms of a binary model. If, say, a three-component stellar model were adopted, there would be more parameters to manipulate (Bahcall *et al* 1974), and it may be possible to avoid a massive compact object. Standard theories predict that when the collapse of a stellar core with a mass of more than two solar masses occurs, a black hole will be formed, and it appears that in Cyg X-1 we are seeing a system in which this has occurred.

Circinus X-1 is another x-ray source in the Galaxy which has properties rather similar to those of Cyg X-1, and it may also contain a black hole.

4.3.3. Cygnus X-3

Cygnus X-3 was first noted in x-rays during a scan of the Cygnus region by a rocket-borne detector in 1966 (Giacconi *et al* 1967). At the time, the only feature which stood out was a prominent low-energy cut-off in the x-ray spectrum, which was interpreted as the effect of interstellar absorption on a distant Galactic source.

Subsequent studies of the spectrum of Cyg X-3 by Bleach *et al* (1972) and Parsignault *et al* (1972) have confirmed the presence of this low-energy cut-off. Serlemitsos *et al* (1975) found a significant variation in the spectral shape in two rocket observations made one year apart; the spectrum changed from a smooth black-body shape to a much harder profile with prominent iron line emission. A narrow iron line was also seen by Kestenbaum *et al* (1977) using a Bragg crystal spectrometer.

The x-ray source was found to show a periodic 4·8 hour variation in intensity by Parsignault *et al* (1972) and Sanford and Hawkins (1972). This periodic behaviour differs from that of Cen X-3 and Her X-1 in that it does not represent a complete eclipse; in Cyg X-3 the minimum x-ray intensity is nearly half of that at maximum. A periodic behaviour over 17·75 days has also been observed (Holt *et al* 1976). It is not absolutely clear which of these is the binary period, because the 4·8 hour period could conceivably be that of a slowly rotating magnetic neutron star.

Gamma-rays with energies greater than 35 MeV have been detected from Cyg X-3 by Lamb *et al* (1977). Although the positional accuracy of gamma-ray measurements is very poor, the identification was made certain by the 4·8 hour modulation of the gamma-rays.

Cyg X-3 has not been detected in the visible part of the spectrum, but it has been picked up as a faint infrared star at wavelengths of 1·6 and 2·2 μm by Becklin *et al* (1972). At these wavelengths, the intensity of the star corresponds to that of an 11th or 12th magnitude visible star. Stars down to a magnitude of about 23 can normally be detected by visible-wavelength astronomy. The usual reason for a star to appear very much brighter in the 1·6 and 2·2 μm bands is that the visible light is obscured by interstellar dust. Obscuration is common for stars like Cyg X-3, which lie in the plane of the Milky Way at a distance of 2 kpc or more. On the basis of this argument, the infrared intensity, coupled with the non-observation in the visible, is consistent with that expected from a blue supergiant star at a distance of about 10 kpc. This distance is in agreement with the distance of the radio source as determined from a study of the 21 cm hydrogen

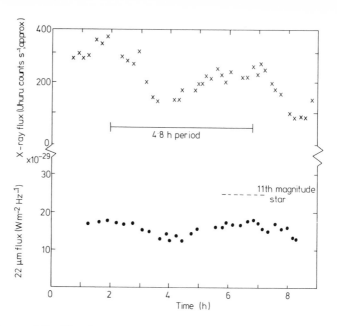

Figure 4.8. Simultaneous x-ray and infrared light curves of Cyg X-3.

line (Chu and Beiging 1973). The infrared observers (Becklin *et al* 1973) went on to study the variability of Cyg X-3, and they found that it exhibited variations in intensity with the same 4·8 hour period as the x-radiation, and with the same phase. This fully confirmed the identification with the x-ray star.

A most interesting feature of Cyg X-3 was accidentally discovered by the radio astronomers. Following the discovery that Sco X-1 showed faint radio emission, searches were made for radio counterparts to other Galactic x-ray sources. In the case of Cyg X-3, an identification was made in 1972 by Braes and Miley (1973). This provided the exact position used for the infrared identification. The weak radio emission (a few tenths of a flux unit) suggested that Cyg X-3 fitted into the group of Galactic sources Sco X-1, Cyg X-2, Cyg X-1 and GX 17 +2, which also showed weak radio emission. Quite by chance, on the evening of 2 September 1972, Gregory *et al* (1972) made a brief observation of Cyg X-3 before moving on to other work. They measured a flux of about 20 flux units, or 100 times the normal intensity of the

86

source. Like the other x-ray source counterparts, some degree of variability was a normal characteristic, but never had such a large change been observed previously. Within hours, two further radio observatories were monitoring the source, and they fully confirmed the existence of the flare. The duration of the flare amounted to only a few days, although at the much lower frequency of 408 MHz (the initial observation had been made at $10\cdot5$ GHz $= 10\,500$ MHz) the flare peak was delayed by nearly two days.

Monitoring of Cyg X-3 in the radio region was continued by several groups through the weeks following the flare-up, and a second sequence of flares was seen to commence 17 days after the initial outburst. This time, three individual maxima were recorded at high radio frequencies at intervals of about three days, although the maximum intensities did not reach the earlier peak values. Again the flare, as seen at lower radio frequencies, lagged behind that at the higher frequencies. The number of observers involved in studying these radio flares is too great to refer to individually here, and the interested reader is referred to a review by Hjellming (1973). Although the events appear highly dramatic to the radio astronomer, it should be borne in mind that the total integrated energy output in the radio spectrum from the first

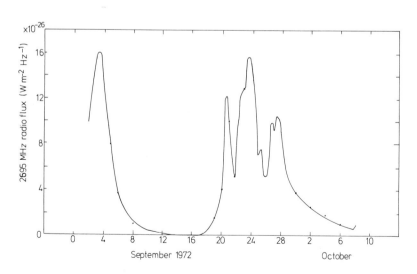

Figure 4.9. The first observed radio flare of Cyg X-3. Data from Hjellming (1973).

flare only amounted to the normal x-ray output over 100 seconds (Canizares *et al* 1973). The radio burst emission has been generally interpreted as synchrotron radiation from high-energy electrons interacting with a magnetic field around the compact component of the binary system. A subsequent radio outburst has been reported by Dashido *et al* (1974).

X-ray observations made before, during and after the first radio outburst (Parsignault *et al* 1972, Conner *et al* 1972) showed nothing unusual, apart from the random variability which is a common feature of all the Galactic x-ray stars.

Cygnus X-3 is most probably an accreting binary x-ray system rather like the others discussed in §§ 4.2 and 4.3. The lack of any observations of Doppler-shifted pulsations or spectral lines makes Cyg X-3 somewhat more mysterious than the other sources which have been described. So far as is known at the time of writing, the gamma-ray emission and the radio flares make Cyg X-3 unique. It is possible that the two phenomena could be related, because the relativistic electrons which are needed to produce the radio emission, could also produce the gamma-radiation by inverse Compton scattering of the x-ray photons.

4.4. X-ray Transients

X-ray transients, or x-ray novae, are stars which show a dramatic rise in x-ray intensity from a previous low (or undetected) state, and then decay away. The rise usually occurs on a timescale of a few days or less, whilst the fall takes weeks or months. This behaviour is rather similar to that of a normal nova at visible wavelengths. However, not all visible novae are seen in x-rays—Nova Cygni was very bright at visible wavelengths in 1975, but was definitely not detected in x-rays (Hoffman *et al* 1976).

The first x-ray transient, Cen X-2, was reported by Chodil *et al* (1967) who analysed the data from a series of rocket shots. Attention was only seriously concentrated on transient sources after a number had been detected by the instruments on board the Ariel V satellite. The two transients for which most information is available are A0620−00 and A0535+26, and these will be described.

A0620 −00 reached its peak intensity in August 1975 (Elvis *et al* 1975), at which time it was brighter than any other x-ray source in the sky. Prior to its rise to its maximum intensity, a 'precursor peak' was observed, and this precursor had a harder spectrum than the later development of the transient (Ricketts *et al* 1975). After the precursor, the rise to maximum took about five days in the 2–18 keV band. The decay was much slower, and was traced over $2\frac{1}{2}$ months by the all-sky monitor experiment on Ariel V (Kaluzienski *et al* 1977). A small flare was seen about two months after the maximum.

Soon after the discovery of x-rays from A0620 −00 by the Ariel V workers, its position was measured accurately using the rotating modulation collimator on the SAS-3 satellite (Doxsey *et al* 1976). Boley *et al* (1976) photographed the appropriate region of sky and found a star which had brightened from magnitude 17·5 to magnitude 11·4 in blue light within a few days of the appearance of the x-ray nova. This constituted a certain identification, and further studies of the optical counterpart showed that it decayed away rather more slowly than the x-ray object. Spectrograms of the visible star showed a featureless continuum. Studies of a collection of sky survey plates by Eachus *et al* (1976) revealed that the visible star had undergone a previous outburst in 1917. It appears, therefore, that the object is a recurrent nova. Radio emission was also detected during the x-ray outburst (Owen *et al* 1976, Davis *et al* 1975).

Estimates of the distance to A0620 −00 by optical spectroscopic techniques yielded a distance in excess of 2 kpc (Gull *et al* 1976, Wickramasinghe and Warren 1976). This implies that the x-ray luminosity was greater than 3×10^{38} erg s^{-1} when the source was at its maximum. The high luminosity has led to suggestions that the source may consist of a black hole in a close binary system.

Some hints of a binary star period of about eight days in A0620 −00 were seen by Matilsky *et al* (1976) in x-rays, and by Tsunemi *et al* (1977) in the visible. It is probable that A0620 −00 is a recurrent nova-like system, consisting of a compact star (possibly a black hole to account for the high luminosity) and a low-mass main sequence star in a close

89

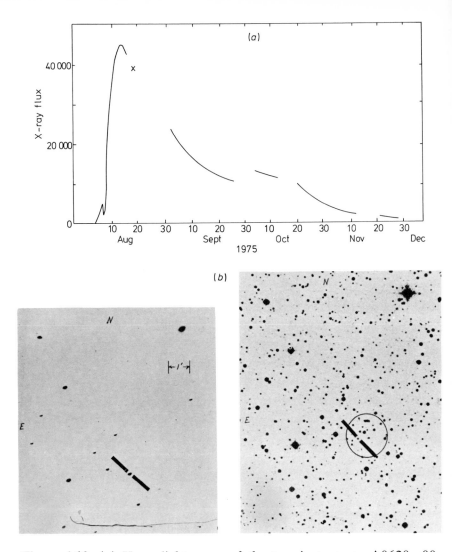

Figure 4.10. (*a*) X-ray light curve of the transient source A0620 −00. Data from Elvis *et al* (1975) and Kaluzienski *et al* (1977). (*b*) The optical identification of the x-ray transient A0620 −00. The better quality photograph on the right was taken in 1955, and that on the left in August 1975. Reproduced from Doxsey *et al* (1976).

binary association. The outburst then consisted of a burst of accretion onto the compact star. What triggered the outburst is something of a mystery. The featureless visible spectrum is just what is to be expected from the accretion disc around the compact star.

The transient A0535 +26 reached its peak in late April

90

1975 and, like A0620 −00, was discovered by the Ariel V workers (Rosenberg *et al* 1975). Like AO620 −00, it showed a precursor, then a rise to its main peak, followed by a slow decline (Kaluzienski *et al* 1975). It also exhibited a small flare some five months after its peak. Spectral measurements made over the first month after the peak showed a hard spectrum whose shape changed very little (Ricketts *et al* 1975).

The most important feature observed in A0535 +26 was the existence of pulsations with a period of 104 seconds. These pulsations were very similar to those observed in the 'slow rotator' eclipsing binary x-ray source Vela X-1 (4U 0900 −40). The pulsations in A0535 +26, discovered by Rosenberg *et al* (1975) showed a Doppler shift in period which was consistent with the presence of a spinning neutron star in a binary system with a period of more than 17 days (Rappaport *et al* 1976). This strongly suggests the similarity between this transient source and the binary x-ray sources.

Several other x-ray transients have been observed with properties rather similar to those of the two described above. A1524 −61 exhibited a precursor peak, rose to its main peak in 20 days, and decayed by a factor of three over two months. A1724 −28 was seen to decay by a factor of three over two

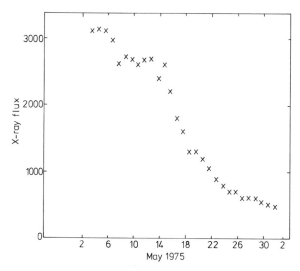

Figure 4.11. The decline of the x-ray transient A0535 +26. Data from Ricketts *et al* (1975).

weeks, and was optically identified. A1118 −61 decayed even more quickly—by a factor of three in a week—and showed pulsations with a period of 6·7 minutes. A number of transients with irregular light curves have been seen. These are probably just normal stellar x-ray sources which exhibit a more marked variability than usual; they are known as 'irregular transients'.

The known transients, of both normal (classical) and irregular types, are closely confined to the Milky Way. It seems likely that they form a sub-class of the bright binary x-ray stars of the Cen X-3 type, although the precise mechanism which triggers their outbursts is unclear.

4.5. X-ray Sources Close to the Galactic Centre

The binary x-ray stars described in §§ 4.2 and 4.3 and the transients described in § 4.4 are closely confined to the plane of the Milky Way, but are spread all around the plane of the Galaxy. Three classes of stellar x-ray source are found predominantly in the region of the Milky Way within 25° of the Galactic centre. These are the bright Galactic bulge sources, the globular cluster sources, and the burst sources. Because of their restricted distribution in the sky, it is believed that these sources lie in the Galactic bulge, which is a region of 4 or 5 kpc radius centred on the Galactic centre. It follows that they must lie at distances between 5 and 15 kpc from us. There is evidence from optical astronomy to indicate that the bulge region of the Galaxy is made up mainly of stars which are older than those found in the spiral arms. It is therefore not surprising that the x-ray sources found there differ from those found in the spiral arms.

4.5.1. The Bright Galactic Bulge Sources

The bright Galactic bulge sources are evident as a cluster of bright x-ray objects around the centre of the Galaxy. They all show an irregular variability in their x-ray intensity, but no periodic eclipsing behaviour. The prominent low-energy cut-off in their spectra confirms that they lie at distances of order 10 kpc in the Galactic plane. They have x-ray luminosities of rather more than 10^{38} ergs^{-1}. Very little else can be said

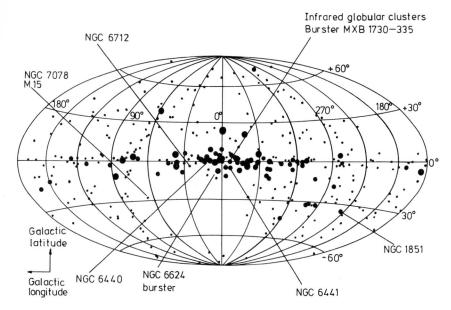

Figure 4.12. A map of the x-ray sky plotted in Galactic coordinates, showing the concentration of strong sources in the bulge. Globular cluster sources are indicated.

about their x-ray properties at present, which is surprising considering that they are amongst the brightest sources in the sky.

Wilson *et al* (1977) have produced a list of accurate positions for the bright sources using the rotating modulation collimator on the Ariel V satellite, and the SAS-3 group have also measured such accurate positions. Two sources GX3+1 and GX9+1 have had their positions measured to arc second accuracy using the lunar occultation technique (Janes *et al* 1972, Davison and Morrison 1977). Optical identification of the x-ray sources has proved to be rather difficult. This is to be expected, because the obscuration by interstellar dust is very heavy for distant stars in that part of the Milky Way.

There is therefore no direct evidence to link the bright Galactic bulge sources with accreting binary systems. That they are so linked is argued from their high luminosities, which correspond to the Eddington limit for compact stars in the mass range 1–5 solar masses. It is hard to envisage how a compact object could accrete sufficient material to produce an Eddington-limited x-ray source other than from a binary companion. The absence of observed eclipses in all the sources places limits on the geometry of the binary systems.

4.5.2. *Globular Cluster Sources*

Some half a dozen x-ray sources are known which coincide in position with globular clusters (Clark *et al* 1976, Ulmer *et al* 1976). One of these sources, MXB 1730−335, coincides with an obscured globular cluster which can only be clearly seen in the infrared.

Globular clusters are spherically shaped condensations, each containing around one hundred thousand stars. Most appear to be made of Population II stars, which have spectra indicating that they are deficient in elements heavier than helium. The globular clusters move in elliptical orbits around the centre of the Galaxy, and these orbits are not confined to the plane of the Milky Way. It is generally believed that the globular clusters were formed very early in the life of the Galaxy, before the elements heavier than helium had been synthesised in stellar interiors and ejected into the interstellar medium by supernova explosions. Stars are much more closely packed in globular clusters than in the spiral arms of the Galaxy. About 120 globulars have been identified in our Galaxy.

The x-ray sources in the globular clusters NGC 6440, NGC 6441 and NGC 7078 (Ulmer *et al* 1976) are rather like the sources described in § 4.5.1—they show random variability but no eclipsing binary behaviour. The sources in NGC 6624, NGC 1851 and in the obscured globular cluster are burst sources (see § 4.5.3).

It is generally assumed that the x-ray sources are to be found at the centres of the globular clusters concerned, simply because that is where the highest density of stars is found. The nature of the x-ray sources is unclear. Clark *et al* (1975) regard them as normal accreting binary star systems, whereas Bahcall and Ostriker (1975) argue in favour of accretion onto a black hole at the cluster centre. There is some evidence that the x-ray globulars have a bright nucleus in the visible.

4.5.3. X-ray Burst Sources

The first x-ray burster was announced by Grindlay and Gursky (1976). A flurry of publications followed, and several dozen bursters are now known. They are x-ray sources which flare up in a few seconds. Typically, they take about one second to reach their peak, and then die down over the next ten seconds or so. They therefore operate on a timescale a hundred thousand times faster than that of the transient x-ray sources.

Most of the known burst sources exhibit bursts which repeat themselves at intervals of hours or days. This recurrence is not exactly periodic in any of the sources; a burst may occur earlier or later than would be predicted by a strictly periodic law. It is therefore not possible to associate the burst interval with the period of a binary system. The most regular burst intervals have been observed from the source MXB 1659 −29 (Lewin and Joss 1977). Bursts from this source recur with an interval which slowly varies between 2·0 and 2·6 hours. A series of 17 consecutive bursts was observed, in which the individual bursts fluctuated by about five minutes in arrival time from a strictly periodic law. Only one burst has been seen from some of the burster sources; this may be because observational coverage has not been adequate to see bursts which recur on a timescale of weeks or more. The 'rapid burster', MXB 1730 −335 (Lewin *et al* 1976, Grindlay and Gursky 1976), exhibits bursts which can recur as fast as once every six seconds, much faster than from any other burster known.

The profiles of intensity against time of the bursts vary from source to source. Some show a single-peak pulse, whereas others have more complex profiles. For example, MXB 1743 −29 has a consistent double- or triple-peak burst profile. The x-ray spectrum usually varies throughout a burst, usually becoming softer as the burst decays. The behaviour of the rapid burster is interesting; the burst amplitude appears to be roughly proportional to the time interval before the next burst. This is rather like the behaviour of a relaxation oscillator.

Many of the bursters appear to be associated with steady

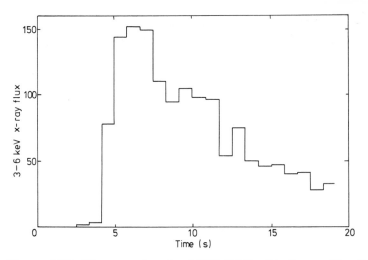

Figure 4.13. The x-ray burster MXB 1728 −34: the profile of a typical burst. Data from Hoffman *et al* (1976).

x-ray sources (Carpenter *et al* 1976). MXB 0512 −40 and MXB 1820 −20 are associated with the steady x-ray sources in the globular clusters NGC 1851 and NGC 6624, respectively (Forman and Jones 1976, Clark *et al* 1976). The rapid burster has no steady x-ray counterpart, but does seem to lie in the obscured globular cluster detected in the near infrared by Liller (1977). Some of the bursters with steady x-ray counterparts whose positions have been determined accurately seem to coincide with blue stars.

Since most of the bursters are concentrated in the plane of the Milky Way and lie within 30 or 40° of the Galactic centre, it is reasonable to suppose that they are clustered around the Galactic centre at a distance of about 10 kpc. At peak, they are about as bright as the strong Galactic bulge sources as described in § 4.5.1, and they therefore have peak luminosities of order 10^{38} erg s^{-1}.

The mechanism which produces x-ray bursts is at present uncertain. One possibility is that rapid fluctuations in accretion rate occur in a normal x-ray binary. Another possibility is that flashes of nuclear fusion occur on the surface of the neutron star in an accreting x-ray binary. Both of these mechanisms could produce the observed luminosity. There is as yet no direct evidence that bursters are actually situated in binary systems.

96

4.6. Low-Luminosity X-ray Stars

A few x-ray sources are known which have been identified with relatively nearby stars. AM Her is an eclipsing binary x-ray source which has been studied in some detail. It has a period of 3·1 hours, and has been identified with the variable star from which it takes its name. It almost certainly consists of a low-mass star in association with a white dwarf. SS Cyg and U Gem are dwarf novae and are seen at x-ray wavelengths. These are probably also binary systems containing white dwarf components. Ultra-soft x-rays have been detected from some very nearby bright stars, namely Capella, Sirius and Algol. Possibly these stars have coronas which emit x-rays much more intensely than does the Sun.

5. *Supernova Remnants*

5.1. Introduction

Several x-ray sources have been definitely identified with supernova remnants within our Galaxy. Supernova remnants (SNR) were well studied before the days of x-ray astronomy; accordingly their visible and radio wavelength properties will be discussed first.

At radio wavelengths, SNR are prominent objects which are seen concentrated towards the Galactic plane. They are diffuse objects with non-thermal spectra (the radio flux density decreases with increasing radio frequency). The majority of SNR show a shell-like spatial structure. In the visible part of the spectrum, faint wisps emitting high-excitation spectral lines can be detected, and there is evidence that these gaseous regions are moving away from the centre of the remnant at high velocities. The radio emission is generally believed to arise from the synchrotron radiation of relativistic electrons in magnetic fields, whilst the visible radiation comes from regions of high-temperature gas.

The association of these diffuse objects with violently exploding stars, or supernovae, is based on historical evidence. In 1572, Tycho Brahe observed what we now know to have been a supernova in the constellation of Cassiopeia; and in 1604, Kepler saw a similar bright, star-like object appear in Ophiucus. Radio supernova remnants are now seen in those positions. Unfortunately, supernova explosions in our Galaxy are rare events, and none has been seen since the seventeenth century. However, the Chinese and the Koreans were in the habit of keeping records of unusual happenings in the heavens before the re-awakening of interest in Europe and, by searching their records, it has been possible to identify some other remnants with supernova explosions which were recorded. The most notable of these is the Crab

Nebula in Taurus, which corresponds to a supernova explosion observed in AD 1054. This must have been a remarkable event to have witnessed, since it was visible in full daylight.

Having mentioned the Crab Nebula, it should be said immediately that although it is one of the most remarkable objects known to astronomers, it is not entirely typical of supernova remnants. Accordingly, it will be discussed separately from the other SNR which emit x-rays. The detection of x-rays from SNR was not altogether unexpected, and has not drastically changed our understanding of these objects. Before the x-ray observations are discussed, therefore, § 5.2 will review our general understanding of how supernova explosions occur, and how the remnants are formed.

5.2. The Formation of Supernova Remnants

Supernova explosions are rare events; they are thought to occur with a frequency of something between 3 and 20 per thousand years in our own Galaxy, but we cannot expect to observe more than a fraction of these in the visible part of the spectrum because of the obscuring effect of the interstellar dust. There is strong evidence that the powerful radio-emitting remnant Cas A corresponds to an explosion in about AD 1670, and that interstellar obscuration prevented its observation. Studies of supernova explosions must therefore be made on supernovae which occur in external galaxies. External galaxies are surveyed regularly to search for new supernova explosions, and about 20 are found each year. Most are so distant as to make detailed studies impossible.

Supernovae are distinguished from ordinary novae in that they have a much greater intrinsic luminosity in the visible. Supernovae at maximum light have absolute magnitudes in the range −16 to −20, and so can be seen quite readily against the background galaxies (which normally have absolute magnitudes in the range −16 to −22). Novae have absolute magnitudes at maximum of only about −7. A nova explosion appears not to disrupt seriously the two stars involved, whereas a supernova drastically damages the original star.

The light curve of a supernova rises to a maximum over a period of ten days or so. It then decays by two or three magnitudes for a month. This is followed by a much slower decay over the next few months. Supernovae have conventionally been classified as 'type I' or 'type II'. After the initial fast decay, type I supernovae decay almost linearly in magnitude, whereas type II light curves flatten off for about 50 days, then decay away rapidly. Type I supernovae are brighter than their type II counterparts, and there are clear spectral differences between the two classes of supernova. It has been common practice to associate type II events with the explosion of massive Population I stars, and type I events with the explosion of older Population II stars (an association which typifies the contrariness of astronomical terminology). Recent workers prefer to classify supernovae into more than these two simple categories.

The optical spectra of supernovae are not easy to interpret. The one point that is clear is that the lines present are severely broadened due to Doppler shifts. This broadening indicates shells of gas moving at velocities in the range $7000-20\,000$ km s^{-1}. Many of the lines are unidentified, and overlap to such an extent that it is difficult to decide which is an emission and which is an absorption feature. If the majority of the features are regarded as being absorptions, then the underlying continuum has a black-body form with a temperature of about $10\,000$ K at maximum, falling to about 5000 K during the first month.

The x-ray detection of a supernova explosion would be exciting. No such detection has been made to date, and the upper limits obtained—of order 10^{-11} erg cm^{-2} s^{-1}—are not of great significance considering the great distance of extragalactic supernovae.

The behaviour of a supernova outburst has been outlined by Woltjer (1972, 1974). He discusses a type II outburst involving a star of 10 solar masses or more. As this rather massive star evolves, the interior passes through phases of nuclear burning at successively increasing temperatures. Eventually instability sets in, resulting in the collapse of the core of the star. This collapse occurs with such violence that the outer layers of the star are thrown off at nearly relativistic velocities. Probably a neutron star is left behind. Spectroscopic studies of the ejected material suggest that about 0·3 solar masses of stellar material leaves with a velocity of approximately 10 000 km s^{-1}, followed by about 5 solar masses at approximately 2000 km s^{-1}. The total kinetic energy of the ejected material is of order 10^{51} erg, which is equivalent to the total energy radiated by the Sun during its lifetime.

The subsequent formation of the supernova remnant is expected to occur in the following way.

Phase I: For the first 90 years after the outburst, the expelled material will move outwards at about 10 000 km s^{-1}.

Phase II: For the next 22 000 years or so, the outward-moving material will collect significant amounts of interstellar gas, so that the remnant will significantly increase its mass. Radiative energy losses will still be unimportant, so that the expansion energy will be shared between the original and the swept up material in the remnant. The expansion velocity will therefore diminish as more material is swept up. This phase will persist until the expansion velocity drops to about 200 km s^{-1}.

Phase III: Once the expansion velocity has dropped sufficiently, the region behind the shock front, separating the expanding remnant from the interstellar medium, will cool to a temperature below 5×10^6 K. Radiative energy losses then become important because the heavier elements present are no longer fully ionised. The shell now moves with constant radial momentum. This phase persists for perhaps 750 000 years.

Phase IV: The expansion velocity has dropped to about 10 km s^{-1}, which equals the random velocity of the interstellar medium. The supernova remnant loses its identity.

The SNR which are observed in the radio, visible and x-ray spectral regions are mostly in phase II of their development. This is sometimes called the 'adiabatic' phase. Cas A and Tycho are very young, and are probably only at the beginning of the adiabatic phase.

The above discussion is based on the hydrodynamics of a mass of gas ejected into a uniform interstellar medium. Two complications need to be borne in mind. Firstly, the interstellar medium is not very uniform, which probably accounts for the gaps in the shell structure of several supernova remnants—the Cygnus Loop for example. Secondly, no account has been taken of the magnetic field and cosmic ray particles which produce the observed synchrotron radio emission. In the case of the Crab Nebula, this magnetic field has sufficient pressure to cause the nebula to accelerate in its expansion.

It is not yet clear whether a central pulsar (rotating magnetic neutron star) is a general feature of supernova remnants. Pulsars have so far only been detected in the Crab and Vela X remnants, but other remnants might well possess pulsars which have not been detected—either because they are too faint, or because their radiation is not beamed towards the Solar System. The idea that every supernova begets a pulsar is attractive. The initial collapse of the pre-supernova star can produce the required neutron star; the rotation of the neutron star, needed to produce the pulsations, comes from the angular momentum of the original star; and the necessary large magnetic field results from the compression of the magnetic field of the pre-supernova star.

There are therefore two reasons why supernova remnants are expected to emit x-rays. Thermal emission is expected from the hot gas behind the shock front, which is at a temperature of more than 5×10^6 K in remnants in the adiabatic expansion phase. If cosmic rays with sufficient energy are present, then synchrotron x-rays will be generated.

5.3. X-ray Supernova Remnants

5.3.1. *Cassiopeia A and Tycho's* SNR

Cassiopeia A (3C 461) is the brightest radio source in the sky outside the Solar System. It has a radio diameter of about 4 arc minutes, and shows a broken, but clearly shell-like structure (Hogg 1969, Rosenberg *et al* 1970). The radio emission is polarised; as with many other radio sources, this is strong evidence for the synchrotron origin of the radiation. The spectral index is 0·76. In the visible, the remnant is not

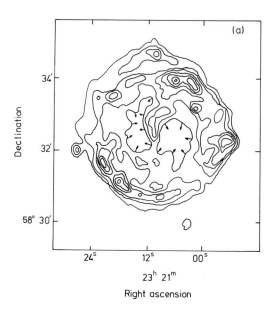

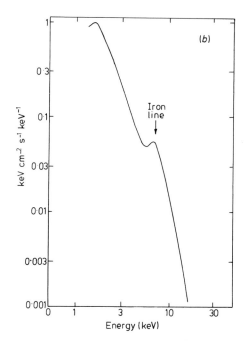

Figure 5.1. (*a*) A 2·7 GHz radio map of Cas A. The arrowed contours represent regions of low emission. (*b*) The x-ray spectrum of Cas A. Data from Davison *et al* (1976).

103

very prominent; both stationary and moving emission knots can be seen, but the moving knots have lifetimes of only about a decade (Van den Bergh and Dodd 1970). The proper motions of the moving knots range from 0·2 to 0·5 arc seconds per year, and their radial velocities average about 5500 km s^{-1}; from this, Van den Bergh (1971) has deduced an approximate distance of 2·8 kpc to the source. Some 21 cm absorption line studies of the radio source (Hagen *et al* 1955) suggest that the object is rather more than 3 kpc distant. Extrapolating back the observed proper motion of the optical features suggests that the supernova exploded about 300 years ago; interstellar extinction could explain why the event was not observed. The idea that this supernova remnant is young is further supported by an observed decrease in the radio flux of 1·1% per year (Mayer *et al* 1965).

Cas A was first detected as an x-ray source by Friedman *et al* (1967), although at that time the accuracy of the position obtained was not sufficient to make the identification certain. The identification was made definite by Gorenstein *et al* (1970), who measured the position using a sounding rocket, and by Giacconi *et al* (1972) using the Uhuru satellite. The 1–10 keV intensity of the source corresponds to about 0·1 of that of the Crab Nebula, though the spectrum was recognised early on to be softer than that of the Crab. More recent studies of the spectrum clearly indicate a thermal origin for the x-ray spectrum of Cas A. The definitive Ariel V data of Davison *et al* (1976) required a two component bremsstrahlung fit, involving regions at 12×10^6 and 60×10^6 K. This spectrum showed clear evidence of iron line emission at 6·7 keV. The presence of these iron lines was confirmed by Pravdo *et al* (1976).

The structure of Cas A was studied using a 3 arc minute beam with the Copernicus satellite. The data indicate that the x-ray emission comes from a region of about the same extent as the radio emission (Fabian *et al* 1973, Charles *et al* 1977).

Since Cas A is a young supernova remnant, and is a very strong source in the radio region, it seems a likely candidate in which to search for a central pulsar; for theoretical reasons one suspects that supernova explosions should leave behind a pulsar. All such searches for a pulsar at radio, optical and x-ray frequencies (Holt *et al* 1973) have yielded null results. This is discouraging, but does not entirely prove that a pulsar is not present, since its radiation may be beamed away from the Sun.

Tycho's supernova remnant is rather similar in its radio and x-ray properties to Cas A.

5.3.2. *Puppis A, Vela X and the Cygnus Loop*

These three supernova remnants are much larger than Cas A, Tycho and the Crab, and are believed to be much older. Accordingly there are no records of the outbursts of the supernova explosions responsible.

Puppis A has a radio size of about one degree in the sky. The distance is rather uncertain, but Palmieri *et al* (1971) quote 1·2 kpc; this suggests an extent of 20 pc for the remnant. The radio source is shell-like, and shows a bright concentration on its eastern side. Optical studies by Baade and Minkowski (1955) have revealed the presence of filaments with a velocity dispersion of 150–200 km s^{-1}. This is much lower than the velocity seen in the optical filaments of Cas A, and is as expected for a supernova remnant much older than Cas A.

X-rays from Pup A were first identified by Palmieri *et al* (1971). Pup A and the Vela X remnant lie close together in the sky, so that earlier observers had confused the two sources under the name Vela X-2. The identification of the Pup A x-ray source was confirmed by Seward *et al* (1971) who, like Palmieri *et al*, also used a rocket-borne instrument. Being a source of soft x-rays, Pup A was much better studied with a thin-window proportional counter on a rocket, than with a satellite such as Uhuru which was only sensitive to x-ray energies above 2 keV.

Spectral data from three rocket flights sensitive over the energy range 0·2–18 keV were analysed by Burginyon *et al* (1975). They found that power-law, exponential and black-body models all yielded unacceptable fits. However, thermal plasmas at temperatures below 10^7 K are expected to produce spectra which depart markedly from an exponential form owing to the presence of strong emission lines. Burginyon *et al* were able to fit their observed spectrum, which had a strong peak at 0·65 keV, to that for a thermal plasma at 4×10^6 K. They used the calculations of Tucker and Koren (1971), and adjusted the elemental abundance to optimise their fit. (It was, as usual, necessary to assume some inter-stellar absorption along the line of sight.) More recent spectral measurements are in approximate agreement; Moore and Garmire (1976) flew a rocket instrument sensitive to x-rays down to 0·1 keV and obtained a temperature of $3 \cdot 6 \times 10^6$ K. The argument for the x-rays being thermal in origin was strongly reinforced when Zarnecki and Culhane (1977) ob-served an oxygen spectral line at 0·66 keV using a Bragg crystal spectrometer.

The spatial structure of Puppis A has been investigated by Zarnecki *et al* (1973) from the Copernicus satellite, and by Catura and Acton (1976) from a rocket-borne low-energy collector. The x-ray source is of about the same size as the radio source, but the region of peak x-ray brightness is displaced from the area of maximum radio brightness.

Vela X has the largest diameter of the bright supernova remnants; it extends over 7 or 8 degrees in the sky. It is among the brightest and most highly polarised of the radio sources in the sky (Milne 1968). In the visible, a bright D-shaped collection of filaments can be seen in the red H-alpha light of hydrogen. The distance to Vela X is only about 500 pc. It is generally believed that the supernova remnant lies towards the centre of the much more extended visible structure known as the Gum Nebula (Brandt *et al* 1976). The great size of the supernova remnant suggests an age of several thousand years or more.

Like Puppis A, Vela X is a source of soft x-rays better studied with low-energy, thin-window proportional counters. At 1 keV, Vela X is the strongest x-ray source in the sky, apart from the Sun. Its spectrum is consistent with that of a hot gas at a temperature of about 4×10^6 K (Moore and Garmire 1976, Clark and Culhane 1976). Its x-ray structure has been studied most recently by Moore and Garmire (1976); it has a patchy appearance, is brightest in its north-western corner, and shows the shell-like structure so charac-teristic of radio maps of supernova remnants.

Vela X is notable in that it is the only supernova remnant

other than the Crab Nebula in which a central pulsar has been detected. The pulsar PSR 0833 −45 is seen at radio wavelengths, and has a period of 87 ms. It has also been very weakly detected at optical wavelengths (Wallace *et al* 1977) and at gamma-ray wavelengths (Thomson *et al* 1977). It now seems clear that, despite an early report to the contrary, no pulsations from this pulsar can be seen in x-rays. There does however seem to be an unpulsed x-ray source in the direction of PSR 0833 −45.

The Cygnus Loop is another of the more extended, older supernova remnants which emit soft x-rays. This remnant is known to optical astronomers as the Veil Nebula.

5.3.3. *Other X-ray Emitting Supernova Remnants*

A few further supernova remnants have been detected in x-rays, but they have not been studied in any detail. Amongst them are IC 443, SN 1006 and MSH 14 −63. Ultra-soft x-rays have been detected from the North Polar Spur radio source, which is thought to be a very old supernova remnant.

5.4. The Crab Nebula

The Crab Nebula is the first object in Messier's catalogue of diffuse visible objects which was compiled in the middle of the eighteenth century, although it is by no means as easy to spot with a small telescope as is M31—the Andromeda Nebula. The name 'Crab' was suggested from its shape in a drawing made by Lord Rosse in the nineteenth century. Today the nebula is known by the following names: M1 (Messier number), NGC 1952 (New General Catalogue of diffuse visible objects), Taurus A (radio astronomy designation), 3C 144 (third Cambridge catalogue of radio sources), Tau X-1 (x-ray astronomy designation), and SN 1054. The last designation refers to the supernova of AD 1054 which was recorded in Chinese and Japanese records, and with which the Crab remnant is certainly associated. When it was observed, SN 1054 appeared as a 'guest star' brighter than Jupiter.

The visible spectrum of the Crab is unusual in that it consists of emission lines superimposed upon continuum radiation. The emission lines are those expected from a hot gas of low density, and are seen from other supernova remnants, and from gaseous nebulae such as the planetary nebulae. The continuum radiation, on the other hand, is not seen from other gaseous nebulae. It is highly polarised and is attributed to a synchrotron emission mechanism. It is this visible synchrotron radiation which gives the Crab its amorphous appearance, and which sets the Crab apart from the other supernova remnants which are seen in the visible.

When photographed through a filter which selects the emission lines and rejects the synchrotron continuum, the Crab shows a filamentary appearance. These filaments lend themselves to two types of observation: (i) because they are

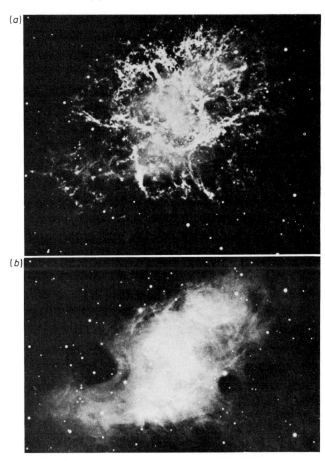

Figure 5.2. Photographs of the Crab Nebula. (*a*) The filaments seen in line emission, and (*b*) the amorphous structure seen in continuum synchrotron emission.

clearly discernible features on a photograph, their proper motions can be measured; and (ii) because they have sharp line spectra, their Doppler shifts can be measured.

The proper motion data indicate that the filamentary structure is expanding away from a point in the centre of the nebula, as is to be expected from the remnant of an explosive event of some 900 years ago. As the age of the remnant is precisely known, the expected proper motion of each filament can be calculated as a function of its angular distance from the explosion centre, on the assumption of a uniform expansion velocity. Along the major axis of the elliptical nebula, the maximum calculated proper motion amounts to about 0·20 arc seconds per year, but the maximum observed proper motion is about 0·035 arc seconds per year greater than this. This behaviour repeats all over the nebula, implying that the motion of the gaseous filaments has been accelerated since the supernova explosion. Such an accelerated motion is very difficult to explain in terms of an interaction between the expanding filaments and the outlying interstellar medium; rather, the reverse effect would be expected. The acceleration is now convincingly explained as resulting from the pressure of the magnetic field within the nebula.

So far, nothing has been said about the distance of the Crab. From a comparison of the proper motion figures with the Doppler shift figures, the distance has been determined to be about 2 kpc. This makes the linear extent of the nebula some 2 pc.

The Crab Nebula is a prominent radio source known as Taurus A. It differs from the radio sources corresponding to the supernova remnants already discussed in that it does not show a shell-like structure; rather, it shows an approximately uniform intensity distribution which falls off towards the edges. Its envelope has an elliptical form similar to that of the optical continuum radiation, and its size is greater than in the visible. The radio spectrum follows a power-law form with a spectral index of 0·26. Most other supernova remnants have a steeper radio spectrum, that is, a larger spectral index.

The synchrotron mechanism was first proposed by Shklovskij in 1953 to account for the radio emission of the Crab. Synchrotron radiation is generated when high-energy

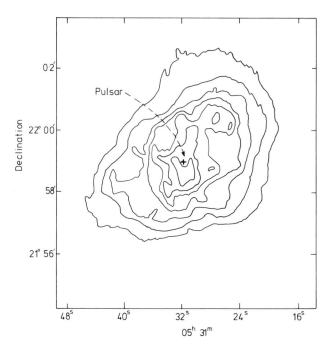

Figure 5.3. A 5 GHz radio contour map of the Crab Nebula.

(relativistic) electrons are made to execute helical motions in a magnetic field (see Chapter 3 for the basic physics). The beauty of the synchrotron mechanism is that it will also account for the optical continuum radiation from the amorphous mass of the nebula, and for the x-radiation to be discussed later. The synchrotron hypothesis leads to the prediction that the radiation should be polarised, with an orientation which correlates with the direction of the magnetic field in the emitting region. Taking the nebula as a whole, the radio polarisation amounts to about 5%, and the visible polarisation to about 9%, the difference being related to the differing sizes of the nebula at radio and visible wavelengths. Over small regions of the nebula, however, the degree of linear polarisation can exceed 50%, and there is a close correspondence between the polarisation patterns at radio and visible wavelengths.

It is possible to estimate the mean value of the magnetic field in the Crab, and thus to calculate the energies and lifetimes of the relativistic electrons producing the synchrotron radiation. By attributing the acceleration of the nebula to

110

magnetic field pressure, a value of about 5×10^{-4} G is obtained for the mean magnetic field. This value is consistent with the observed lack of Faraday rotation in polarised radio signals seen through the nebula, and is also consistent with the low flux of 10^{11} eV gamma-rays due to inverse Compton scattering by the same high-energy electrons (Fazio *et al* 1971). Using this magnetic field value, the electron energies needed to produce the radio emission turn out to be of order 3×10^{8} eV, and their lifetimes are of the same order as the age of the nebula. But production of the optical emission requires electrons with energies of order 2×10^{11} eV, and these have lifetimes against energy loss of only 100 years or so. The lifetimes of x-ray producing electrons are shorter still.

The above calculations have an important consequence for our understanding of the nebula, since they imply that high-energy electrons have been injected much more recently than in the supernova explosion of AD 1054. It was difficult to explain such an injection of electrons until the discovery of the Crab pulsar (NP 0532) in the late 1960s. A spinning magnetic neutron star, which is what a pulsar is recognised to be, has the capability of continuously injecting high-energy electrons into the nebula, and also of building up the strength of the magnetic field. Hence, as is shown in § 5.4.2 describing NP 0532, the pulsar can provide a quantitative explanation of the supply of energy which is radiated by the nebula.

5.4.1. X-rays from the Crab Nebula

The Crab Nebula was the first object to be identified with a celestial x-ray source. Tau X-1 was one of the first x-ray objects to be detected by rocket-borne detectors, and its position coincided roughly with that of the Crab. The technique of lunar occultation was used by Bowyer *et al* (1964) in a brilliant observation which firmly established this identification. The lunar occultation method (which had previously been used to pinpoint the positions of some radio sources) relies on timing the disappearance of an astronomical object as it passes behind the Moon's disc. It is equally useful to time the reappearance of the object as it emerges from eclipse. The position of the Moon is known to an accuracy of better than one second of arc, so that, by sufficiently accurate timing, it is possible to locate source positions with great precision. The Moon moves across the celestial sphere at a rate of about 13 degrees per day, so that a timing accuracy of 1 arc second corresponds to a positional accuracy of some 0·5 arc seconds. Only objects within 5° of the ecliptic are occulted by the Moon. Even for these, occultations occur at monthly intervals for only a few consecutive months, and each of these is visible over only a restricted part of the Earth's surface. The occultation cycle repeats with a period of about 11 years. It was fortunate, therefore, that the Crab should be occulted within a couple of years of the discovery of Tau X-1. The principal difficulty involved in making an x-ray observation of a lunar occultation is the need to launch the rocket at the right time: the four minutes that the rocket is above the atmosphere must overlap the time when the celestial object disappears or reappears. The results of the Crab occultation experiment showed that the x-ray signal from Tau X-1 did disappear at the time when the Crab was passing behind the limb of the Moon, so establishing the identification. The experiment also showed that the x-ray source was extended by about one arc minute, because the x-ray signal was occulted gradually.

The x-ray spectrum of the Crab has been extensively studied, both in the 1–20 keV range by rockets and satellites, and in the 20–500 keV range mainly from balloons. The overall spectrum can be fitted rather well with a single power law. In a careful review of their own measurements and earlier data by other workers, Toor and Seward (1974) found the best fit over the range 2–60 keV to be a power law of the form $I(E) = 9 \cdot 5 \, E^{-1 \cdot 08 \pm 0 \cdot 05} \, \text{keV cm}^{-2} \text{s}^{-1} \text{keV}^{-1}$. In a compilation that included gamma-ray measurements, Parlier et al (1974) found a best fit of $I(E) = 10 \, E^{-1 \cdot 2} \, \text{keV cm}^{-2} \text{s}^{-1} \text{keV}^{-1}$, which suggests that the spectrum steepens slightly at higher energies. The spectrum is so well determined that the Crab is frequently used as a calibration source by x-ray astronomers. It is also the strongest source of x-rays which does not show random variability.

The power-law spectrum fits in neatly with the idea that the x-rays are generated by the synchrotron process, and that

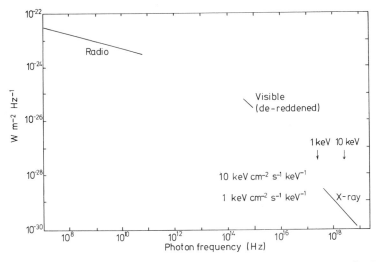

Figure 5.4. The complete electromagnetic spectrum of the Crab Nebula.

the x-ray spectrum reflects the power-law nature of the relativistic electron spectrum. It should be noted, however, that the x-ray spectrum does not fit onto a direct extrapolation of the radio spectrum, and that the slope of the x-ray spectrum is greater than that of the radio spectrum. Radio, optical and x-ray spectra can be fitted together if one supposes that (i) the radio data extrapolate directly to the visible data, and (ii) that just to the infrared side of the visible, the spectrum steepens to a spectral index of 1·1. This then fits the x-ray data in both intensity and slope. A consistent synchrotron model will then account for the radiation in the radio, visible and x-ray regions. The assumption that the power-law electron spectrum steepens at an energy corresponding to infrared radiation is reasonable for a relativistic electron spectrum, because the various energy loss mechanisms which operate on the electrons will tend to steepen the spectrum towards higher energies.

As for the visible and radio emission, the crucial test for the synchrotron mechanism as an explanation of the x-rays is the observation of linear polarisation. This measurement was first made by Novick *et al* (1972), and has been most recently confirmed by Weisskopf *et al* (1978). The observed x-ray polarisation of 18% is considerably larger than that observed for the whole nebula in the radio, and is slightly larger than that seen in the visible. This is quite reasonable when it is remembered that the linear extent of the Crab is smaller in x-rays than in the other spectral regions. By analogy with, say, the visible data, the overall polarisation is a vector sum of many polarised components, each component coming from a separate spatial region. In general, the more components that are viewed at once, the greater will be their tendency to cancel each other out. The x-ray emission, because it extends over a smaller spatial region than the emission at longer wavelengths, will show the smallest degree of cancellation, and might therefore be expected to show a larger overall polarisation.

The spatial extent of the x-radiation from the Crab was explored using the series of lunar occultations which occurred in 1974 and 1975. These were carefully studied by a number of groups using rocket-borne detectors launched from suitably chosen sites on the Earth. Rockets were preferred to satellites for this work because the rapid motion of a satellite round the Earth takes it through the occultation zone too rapidly. The results confirmed the impression formed from the occultation ten years earlier, that the Crab appears smaller in x-rays than in the visible. The later results, which consisted of observations of several disappearances and reappearances, suggested that in x-rays the long axis lies in a NE–SW direction, at right angles to the long axis in the visible and radio. X-ray measurements were published by Ricker *et al* (1975), Davison *et al* (1975), Staubert *et al* (1975), Wolff *et al* (1975), Palmieri *et al* (1975) and Ku *et al* (1976). The variation in the size of the nebula with wavelength can be understood on the basis of the synchrotron model. The higher-energy electrons responsible for the higher-energy radiation have shorter lifetimes and so cannot penetrate to the edges of the nebula. The x-ray size is therefore closer to the size of the electron injection region around the pulsar, which is probably the region in which Scargle (1969) detected moving optical wisps. In the same way, it is possible to explain why the synchrotron visible emission of the nebula comes from a smaller region than the synchrotron radio emission.

The x-ray luminosity of the Crab amounts to some 10^{37} erg s^{-1} in the 0·5–10 keV band. This energy flux is of the same order of magnitude as the visible synchrotron flux, but considerably exceeds the radio flux. If one supposes that the visible and x-ray regions of the spectrum are joined by a single power law, then the total energy emitted by the Crab amounts to about 5×10^{37} erg s^{-1}. This is more than 10 000 times the total energy output of the Sun.

In the foregoing discussion, the x-ray observations have been attributed to the synchrotron radiation of the electrons injected into the nebula by the pulsar. It is by the presence of this highly active pulsar that the Crab differs from the other well studied supernova remnants (although there is a less active pulsar in the Vela X remnant). Faint x-radiation has been observed from a region lying outside the main synchrotron source of the Crab (Toor *et al* 1976, Charles and Culhane 1977). This radiation appears to have a softer spectrum than the stronger synchrotron source, and could well be the same type of thermal radiation that is observed from the expanding shock fronts of other supernova remnants. Perhaps, therefore, the Crab can be pictured as a normal supernova remnant as seen in optical emission lines and soft x-rays, with a stronger synchrotron emission superimposed at all wavelengths by the pulsar-injected electrons.

5.4.2. NP 0532—The Crab Nebula Pulsar

The pulsating star in the Crab Nebula was discovered by radio astronomers (Staelin and Reifenstein 1968). Like the other known radio pulsars, NP 0532 produces radio pulses at a frequency which remains very constant. The period of the pulsations by NP 0532 is $P = 0\cdot0331$ seconds and this is the fastest pulsar known; the majority of pulsars have periods in the range 0·5–2 seconds. The rate of change of the pulsation period, dP/dt, can be measured for NP 0532 and other pulsars by timing pulsations over long periods using sophisticated clocks. For NP 0532 a slowing down is observed, and it corresponds to a change in period of $1\cdot3 \times 10^{-5}$ seconds per year. It is common practice to discuss the 'characteristic age' of a pulsar. It is defined as

$$T = \frac{P}{2\,\mathrm{d}P/\mathrm{d}t},$$

and represents the time taken for the pulsar to halve its frequency if $\mathrm{d}P/\mathrm{d}t$ has remained constant. The characteristic age is generally regarded as giving an order of magnitude estimate of the real age of the pulsar. For the Crab pulsar, the characteristic age is $T = 1300$ years, whereas T is typically a million years for the longer-period pulsars. The characteristic age of NP 0532 therefore supports its identification with the supernova explosion of AD 1054. Other evidence for the identification is that the pulsar lies close to the centre of expansion of the nebula, and that the radio-determined distance of the pulsar is consistent with 2 kpc, which is the optically determined distance of the Crab remnant.

It is necessary at this stage to summarise the theoretical ideas which underlie radio pulsar studies before discussing NP 0532 further. Theorists have attempted to explain all pulsars within the framework of a singe theory. The prime observational features are that the pulses occur with an extreme regularity, and show a very gradual increase in period. (The x-ray objects such as Cen X-3 and Her X-1 discussed in Chapter 4 are different; they show a decrease in period.) Secondary features of pulsars are that the pulse amplitude varies from pulse to pulse, that the pulse profiles tend to be double-peaked, and that radio polarisation varies through the pulse. An additional feature observed in NP 0532 is the occasional occurrence of 'glitches', points in time when small, but permanent decreases in period take place. The extreme regularity of the pulsations requires that a 'clocking' mechanism must be present within the pulsar. It is difficult to explain the observed range of pulsar periods on the basis of a stellar oscillation; accordingly a rotational clocking mechanism has been adopted to explain pulsars. The rotating body is envisaged as being a neutron star with a strong magnetic field. A neutron star is about as massive as the Sun, but has a radius of only some 10 km. Its surface gravity is enormous, and is able to prevent its flying apart even when it is spinning at 30 revolutions a second. The steady slowing down of the 'pulsations' of a pulsar can then be understood as a steady loss of kinetic energy of rotation. The actual production of the radio pulses is attributed to the effect on a surrounding plasma of the strong magnetic field which rotates with the neutron star; this magnetic field must be inclined to the axis of the rotation. It can be seen that the sweeping of a magnetic pole across the line of sight will produce a lighthouse-type effect to account for the radio pulsations. By varying the angle between the line of sight and the spin axis, and the angle between the magnetic and spin axes, it is also possible to explain some of the detailed pulse shape and polarisation observations. One intriguing consequence of this so-called 'oblique rotator' model is that there may well be pulsars whose radiation is beamed away from the Solar System, and which consequently have not been observed.

The energy source in the Crab Nebula can be accounted for in terms of the loss of rotational kinetic energy from NP 0532. If the mass of the neutron star is one solar mass ($= 2 \times 10^{33}$ g), and the radius is 10 km ($= 10^6$ cm), then the moment of inertia of the neutron star will be $I = \frac{2}{5}MR^2 = 8 \times 10^{44}$ CGS units. The kinetic energy of a rotating body is $E = \frac{1}{2}I\omega^2$, where ω is the angular velocity in rad s^{-1}. The rate of change of kinetic energy of a body whose angular velocity is changing at a rate $d\omega/dt$ is therefore $dE/dt = I\omega \, d\omega/dt$. For NP 0532, $\omega = 2\pi/P$, where P is the rotation period, so $\omega = 2\pi/0\cdot0331 = 200$ and $d\omega/dt = -4\pi P^{-2} . dP/dt$, where $dP/dt = 1\cdot3 \times 10^{-5}$ s yr$^{-1} = 4 \times 10^{-13}$; so $d\omega/dt = 5 \times 10^{-9}$. Therefore $dE/dt = I\omega \, d\omega/dt = 8 \times 10^{44} \times 200 \times 5 \times 10^{-9} = 8 \times 10^{38}$ erg s^{-1}. This rate of loss of rotational kinetic energy of the pulsar will be seen to be greater than the observed luminosity of the nebula—5×10^{37} erg s^{-1}. It follows that, purely on energy grounds, the slowing down of the pulsar is more than adequate to explain the continuing energy emission of the nebula. The pulsar is thought to inject energy into the nebula in the form of relativistic electrons and lines of magnetic force. This continuous source of energy overcomes the difficulties in relation to electron lifetime which are encountered in trying to explain the visible and x-ray emission of the nebula in terms of electrons injected at the time of the

117

supernova explosion. The ability of the rotating neutron star theory for the pulsar to explain the luminosity of the nebula in turn lends support to the validity of the rotating neutron star as an explanation for pulsars.

The Crab pulsar can be observed in the visible and x-ray parts of the spectrum, as well as at radio wavelengths. In the visible, it is a weak 16th magnitude object, with an integrated intensity of only one thousandth part of that of the nebula. It is star-like, and possesses a continuous spectrum. As is the case for the radio emission, the pulse profile is double-peaked with the same period of 0·0331 seconds, as is seen in the radio.

At x-ray wavelengths, the pulsar is more prominent. About 10% of the x-radiation is pulsed with the same period as in the radio and visible. The x-ray pulsations were discovered by Fritz *et al* (1969) in the 1–10 keV band. Although these x-ray observations did not precisely locate the position within the nebula from which the pulsations came, the coincidence of the pulsation period with the known radio period was sufficient to establish the identification. The rapidity of the pulsations indicates that the pulsar must be star-like in appearance.

The average x-ray pulse profile has been studied in detail by Rappaport *et al* (1971) and by Fritz *et al* (1971). It shows broadly the same features as are seen in the visible—a sharply peaked 'main pulse' with a broader 'interpulse' some 0·014 seconds later. The interpulse is rather stronger relative to the main pulse in x-rays than it is in visible light; in fact, in the x-ray region the two pulses have roughly the same area under the intensity against time curve The x-ray pulses coincide in absolute arrival time with the optical pulses. The radio pulses, on the other hand, are delayed by their propagation through ionised regions of interstellar space.

The spectrum of the pulsed x-radiation is of a continuous power-law form, and appears to be slightly harder than that of the diffuse nebula. Consequently, the pulsar radiation becomes more prominent at higher x-ray energies. Observations of the pulsar between 20 keV and 1 MeV have been reported by Hillier *et al* (1970) and others, and at gamma-ray energies greater than 1 MeV by Parlier *et al* (1973) and

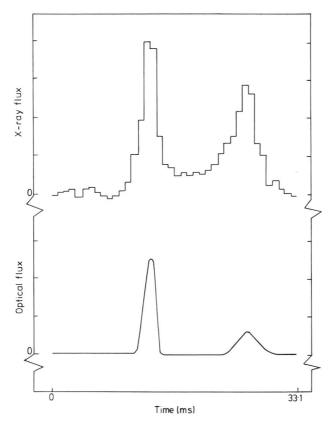

Figure 5.5. The pulsation profile of the Crab pulsar in x-rays and visible light. X-ray data from Rappaport *et al* (1971).

others. The situation, as summarised by Parlier *et al* (1974), is that the x-ray and gamma-ray observations of the pulsar can be fitted to a spectrum of the form $I(E) = 0.6\, E^{-1.05}\,\mathrm{keV\,cm^{-2}\,s^{-1}\,keV^{-1}}$ over the range 10^3–10^9 eV. It appears, therefore, that the x-ray and gamma-ray emissions of the pulsar are parts of the same phenomenon, and that at the highest energies, the radiation from the pulsar at least equals that from the rest of the nebula.

6. Extragalactic X-ray Sources and the Isotropic Background

6.1. Introduction

A large fraction of the 100 or so x-ray sources found at high Galactic latitudes have been identified with extragalactic objects. These objects may be divided into three main categories: (i) nearby galaxies, which contain a number of x-ray sources similar to those in our own Galaxy; (ii) active galaxies, which emit considerably more x-radiation than our own Galaxy; and (iii) clusters of galaxies, where the x-radiation seems to originate in a medium lying between the individual members of the cluster.

The study of extragalactic x-ray sources is less well developed than that of the Galactic sources, because the extragalactic sources appear so faint to present-day instruments. The strongest is 25 times fainter than the Crab, and most are at a level where they can only just be detected by long observations with satellite instruments. The focusing telescope on board the HEAO-B satellite should provide much new information on extragalactic sources.

The fact that distant extragalactic x-ray sources have been detected at all indicates that they must have intrinsic luminosities considerably greater than that of our own Galaxy. We can obtain an estimate of the luminosity of our own Galaxy by adding up the luminosities of the individual sources. Margon and Ostriker (1973) provide a convenient list of the luminosities of the brighter sources which dominate the luminosity of the Galaxy. The total x-ray emission in the 2–10 keV band amounts to just under 2×10^{39} ergs^{-1}. It should be noted that this is much smaller than the visible-

120

wavelength luminosity of the Galaxy—4×10^{45} erg s^{-1}—because, although the x-ray sources are as luminous as very bright stars, there are many more (10^{11}) stars than x-ray sources (perhaps 2×10^2). The limiting sensitivity of satellite instruments, such as Uhuru and the survey instrument on Ariel V, amounts to about 0·002 of the flux of the Crab Nebula, or $S = 4 \times 10^{-11}$ erg cm^{-2} s^{-1}. The maximum distance, D, at which a galaxy like our own could be detected, can be calculated from

$$ S = \frac{L}{4 \pi D^2}, $$

whence $D = 2 \times 10^{24}$ cm $= 0·7$ Mpc. This is approximately the distance to M31, the nearest galaxy similar to our own. All other extragalactic x-ray objects which have been identified (except for the Magellanic Clouds) lie at distances which are considerably greater than this, and must therefore have x-ray luminosities greater than that of our Galaxy. Taking an extreme example, the quasi-stellar object 3C 273 lies at a distance of about 600 Mpc according to current ideas. It yields an x-ray flux at the Earth of about 8×10^{-11} erg cm^{-2} s^{-1}, which means it must have a luminosity of more than 10^{45} erg s^{-1}, a million times the x-ray luminosity of our Galaxy. These high x-ray luminosities point to a fundamental difference between extragalactic x-ray emitters and the x-ray sources found in our Galaxy.

6.2. Normal Galaxies

The Magellanic Clouds are two dwarf galaxies visible to the naked eye from the southern hemisphere of the Earth. They lie at distances of about 55 kpc, only about ten times the distance of the centre of our own Galaxy from the Sun. It is not therefore surprising that it has proved possible to pick out individual stellar sources in both clouds. Four or five such sources have been reported in the Large Cloud (LMC), two of which have been seen to vary in intensity (Tuohy and Rapley 1975). The Small Cloud (SMC) contains three reported sources, the brightest of which, SMC X-1, has been studied in some detail because it shows an eclipsing binary behaviour with a period of 3·89 days (Schreier *et al* 1972). SMC X-1 has been identified with a supergiant BO star by Liller (1973). Those sources which can be picked out in the Large and Small Clouds have similar luminosities (10^{38} erg s^{-1}) to the strongest sources in our Galaxy; so there is also likely to be a population of weaker sources in both clouds. However, since the clouds are both less massive than our Galaxy, one would expect fewer x-ray sources overall.

M31 is the nearest of the spiral galaxies; it can be seen with the naked eye on a clear night from the northern hemisphere. It is possibly twice as massive as our Galaxy, but is otherwise very similar. It is a faint x-ray object, implying an x-ray luminosity of about 2×10^{39} erg s^{-1} (Bowyer *et al* 1974). The x-ray emission appears to be concentrated within the visible-wavelength outline of the galaxy, and the spectrum seems similar to that of the brighter Galactic sources. There is currently no reason to doubt that M31 is populated by x-ray sources similar to those which have been studied in our own Galaxy.

6.3. Active Galaxies

A small fraction of all the galaxies which can be observed with an optical telescope show signs of explosive activity in their nuclei. These galaxies are classified according to the particular way in which the activity manifests itself, but it is currently thought that all active galaxies may be powered by a similar underlying mechanism. The x-ray observations, because of the great luminosities they imply, add considerably to the phenomena which must be accounted for in these unusual galaxies. Physical mechanisms which satisfactorily explain the energy output from Galactic x-ray sources—such as single supernova explosions and mass accretion onto condensed objects of normal stellar mass—fall far short of explaining the x-ray output of active galaxies.

6.3.1. *Radio Galaxies*

Radio galaxies form a well known class of active galaxy. They are galaxies which have a high (greater than 10^{40} erg s^{-1})

radio luminosity compared with galaxies like our own. From radio-wavelength spectral data, it is clear that the emission mechanism is synchrotron radiation involving relativistic electrons moving in a magnetic field. Linear polarisation measurements confirm this. Most, though not all, radio galaxies emit their radio waves from two lobes, which lie at equal distances, and in opposite directions, from the parent galaxy. The total energy needed to fill the lobes with a magnetic field and relativistic electrons typically amounts to 10^{60} erg. This exceeds the energy of a supernova outburst by a factor of at least 10^8. Supernova outbursts are the most violent of well studied Galactic events; hence the difficulty in accounting for the energy input to radio galaxies. It is normally argued that the energy input must originate in the nucleus of the parent galaxy, and the faint, star-like radio sources seen in the nuclei of some radio galaxies probably have some connection with the energy injection.

Centaurus A (NGC 5128) is the radio galaxy which has been most intensively studied at x-ray wavelengths. It lies at a distance of about 4 Mpc, and is the nearest of its class. In the visible it appears as an SO galaxy about 4 minutes of arc in diameter, and is crossed by an obscuring dust lane. Within the dust lane lies a compact infrared source (Becklin *et al* 1971), which is thought to mark the active nucleus. At radio wavelengths, two pairs of lobes are seen. The outer pair are separated by about 8° in the sky, far beyond the visible contours of the galaxy, whilst the inner pair of lobes are only separated by about 7 arc minutes. It is as if the two pairs of lobes had been ejected from the galaxy at different times in the past. Coincident with the nucleus of NGC 5128 is a radio source with an angular diameter of less than 0·001 arc seconds (Price and Stull 1975).

X-rays from Cen A were first detected in a rocket shot by Bowyer *et al* (1970). A subsequent positional determination by Grindlay *et al* (1975) has shown that the x-ray source is associated with the visible galaxy, and not with any of the radio lobes. Two pieces of evidence suggest that the x-ray source is associated with the nucleus of the galaxy. Firstly, like the compact radio source, it is variable in intensity, having changed in x-ray brightness by a factor of three to five

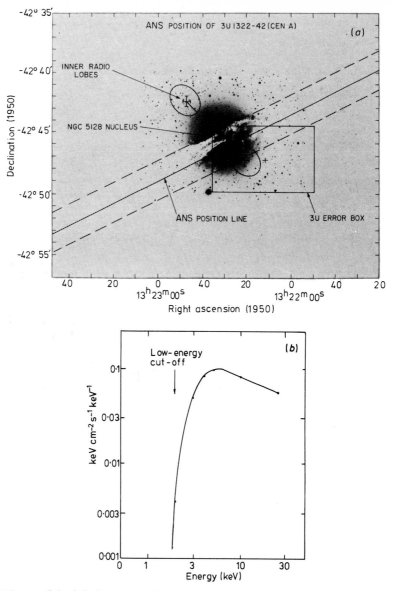

Figure 6.1. (*a*) Some positional measurements on the x-ray source in Cen A, superimposed onto a photograph of the galaxy. From Grindlay *et al* (1975). (*b*) The x-ray spectrum of Cen A. Data from Ives *et al* (1976).

over three or four years (Davison *et al* 1975, Lawrence *et al* 1977). It was even seen to increase in strength by 40% over a six-day period by Winkler and White (1975), which places a severe constraint on the size of the source region. Secondly, the x-ray spectrum shows a severe low-energy cut-off, corresponding to a column density $N_H = 1 \cdot 35 \times 10^{23}$ atoms cm^{-2}. This absorption must be caused by material surrounding the nucleus of NGC 5128 itself. The x-ray spectrum has a power-law form with a spectral index of $0 \cdot 8$ (Stark *et al* 1976). Possible emission mechanisms for this nuclear source are discussed in § 6.3.4.

Cooke *et al* (1978a,b) have detected weak x-ray emission from the giant radio lobes of Cen A. This is probably generated by the inverse Compton interaction between the synchrotron radio photons and the relativistic electrons within the radio lobes.

Cygnus A is the strongest of the radio galaxies, even though it lies at a considerable distance. Like Cen A, it has a double-lobed radio structure. It is probably a weak x-ray source (Kellogg *et al* 1973), though there is a possibility of confusion with a nearby cluster of galaxies.

M87 (Virgo A, NGC 4486) is a peculiar radio galaxy at the centre of the Virgo cluster. Seen in visible light on a short-exposure photograph, it has a jet which appears to have been ejected from the nucleus of the giant elliptical galaxy. Like the Crab Nebula, this jet emits visible synchrotron radiation. The radio emission from the galaxy is strong, but shows a single-lobed structure coincident with the visible galaxy. M87 is probably an x-ray source, but observations are confused by the Virgo cluster x-ray source in which it is immersed (Catura *et al* 1972). Many early reports were made of x-rays from M87 before it was realised that clusters of galaxies also emit x-radiation.

6.3.2. Seyfert Galaxies

Seyfert galaxies are spiral galaxies with bright nuclei. The nuclei show strong and broadened gaseous emission lines in their spectra at visible wavelengths. Such strong emission lines are uncommon in galaxies, most of which show a composite stellar absorption line spectrum. The broadened lines in Seyferts are thought to be generated by gas which has been ejected at high velocity from the nucleus of the galaxy. Both permitted lines from high-density gas, and forbidden lines from low-density gas are present in Seyfert spectra. In type 2 Seyferts, both types of lines have similar widths; in type 1 Seyferts the permitted lines are broader than the forbidden lines, suggesting that the dense central regions of gas are exploding more energetically. The nuclei of Seyfert galaxies of both types are strong infrared sources, and sometimes show variability, implying extremely compact dimensions.

NGC 4151 is a relatively nearby type 1 Seyfert galaxy which was first detected in x-rays by Gursky *et al* (1971). Its x-ray output shows variability (Charles *et al* 1975). Its spectrum (Ives *et al* 1976) is hard and has a power-law form with a spectral index of about 0·6, and shows a prominent low-energy cut-off. NGC 4151 is one of the few extragalactic x-ray sources which has been observed at energies above 25 keV (Auriemma *et al* 1978).

NGC 1275 lies at the centre of the Perseus cluster and is normally classified as a Seyfert galaxy, although it is uncertain whether it is of type 1 or type 2. It is almost certainly an x-ray source (Wolff *et al* 1975), although observations are confused by the presence of the surrounding galaxy cluster emission.

NGC 1068 is the brightest of the Seyferts in the infrared. It is of type 2, and appears not to be an x-ray source.

A further dozen or so type 1 Seyferts have been identified with x-ray sources. Most of these were identified by Elvis *et al* (1978) by comparing the Ariel V catalogue of high Galactic latitude x-ray sources (Cooke *et al* 1978) with catalogues of optically detected Seyfert galaxies.

Type 1 Seyfert galaxies are therefore definitely identified as a class of extragalactic x-ray sources. Variability in x-ray intensity has been observed for a few of these objects, and is probably a general feature of the class. Possible emission mechanisms are discussed in § 6.3.4.

6.3.3. *Quasi-stellar Objects*

Quasi-stellar objects normally appear star-like to the optical observer, but possess peculiar spectra quite unlike those of normal stars. Broad emission lines are seen which are severely redshifted from their normal position. In the early

1960s, the spectra were not understood at all, until it was realised that a large Doppler redshift would allow the lines to be recognised as normal atomic transitions. The redshifts, if interpreted as Doppler shifts, imply velocities of recession of an appreciable fraction of the velocity of light. The standard interpretation for these large velocities is that the QSOs are at great distances and share in the Hubble expansion of the Universe. That they can be observed at all at optical wavelengths implies that they are extremely luminous, and the variability in their light output shows that, like some of the Seyfert galaxies, their energy is radiated from a compact region. They are probably the most extreme example known of the active galaxy family, although evidence to link them to normal galaxies is sparse.

3C 273 is the nearest and brightest of the QSOs, and was first detected by Bowyer *et al* (1970). It has a power-law spectrum with a spectral index of around 0·9, and shows a low-energy cut-off (Culhane 1978). Variability in its x-ray output has also been observed. It therefore resembles the compact x-ray sources in Cen A and the type 1 Seyfert galaxies. Its maximum x-ray luminosity amounts to about 7×10^{45} erg s^{-1}, more than a million times that of our Galaxy.

One or two other weak x-ray sources are possibly identified with QSOs, but they have not been studied in detail.

6.3.4. *The Mechanism of Active Galaxy X-ray Emission*

All the active galaxy x-ray sources discussed in the previous sections are very luminous, have hard spectra and appear to be very compact. Two different theories have been evolved which can account for these features. At present there is insufficient observational evidence to choose between the competing theories.

One theory involves the accretion of material onto a massive (greater than 10^7 solar masses) black hole centrally situated in the active galaxy. The gravitational energy released could heat the gas to a temperature of 10^9 K or so to yield the desired high luminosity and hard spectrum. Fluctuations in accretion rate would account for the variability. Some models of this type can also account for the acceleration of relativistic electrons.

127

The other theory is the so-called Compton synchrotron theory in which radio, infrared and visible radiation is produced in the compact core of the galaxy by synchrotron radiation. Inverse Compton interactions between the relativistic electrons and photons then produce x-ray emission.

6.4. X-ray Emission from Clusters of Galaxies

The clustering of galaxies into groups of up to several thousand members has been known for a long time. Such groups may be seen on the plates of the Palomar sky survey atlas.

Clusters of galaxies are usually centred on giant galaxies, and these dominant central galaxies are often active as well. The Virgo cluster has M87 as its dominant member, whilst the Perseus cluster has NGC 1275 near to its centre. The Coma cluster, on the other hand, is not centred on a galaxy which shows any remarkable activity. Radio galaxies in clusters often show peculiarities; several radio galaxies in Perseus show radio tails (Ryle and Windram 1968) which can be explained by supposing that the radio-emitting region is swept back as the galaxy moves through a 'medium' which pervades the cluster. This 'radio tail' phenomenon is also seen in other clusters. There is evidence for faint optical emission from the intergalactic medium in the Coma cluster (Welch and Sastry 1971), and this is also probably a common property of clusters.

The motions of the galaxies in a cluster may be studied by measuring the Doppler shifts in the spectra of the component galaxies. The motions may be understood as comprising two components: an overall recession which reflects the expansion of the Universe, and a random motion of the galaxies with respect to each other. From the measured random velocities one would have expected the clusters to have dispersed by now, given that the Universe is 10^{10} years old. That they have not dispersed indicates that they must be bound together by a gravitational force. The sum of the masses of the individual galaxies is insufficient to account for the binding force, leading to the suspicion that some form of hidden mass resides somewhere within clusters of galaxies.

The first observation which could be clearly interpreted as x-ray emission from a cluster of galaxies was made when Uhuru detected an extended source in the direction of the Coma cluster (Gursky *et al* 1971). Subsequently, extended x-ray sources have been detected in five other clusters: Perseus, Virgo, Centaurus, Abell 262 and Abell 2256 (Kellogg *et al* 1973). Ironically, x-radiation had been detected from the Virgo cluster several years previously, but it had been attributed to the radio galaxy M87. A two-dimensional x-ray image of the Virgo cluster by Gorenstein and Tucker (1977) shows that M87 is also an x-ray source.

At least 30 further x-ray sources have been identified with clusters (Culhane 1978) with reasonable certainty. Most of these sources are weak and have only poorly determined positions on the sky. Their identification with the clusters in the Abell catalogue is based on the large number of coincidences between x-ray source error boxes and Abell clusters. Only one or two such coincidences would occur by chance if the x-ray sources were un-correlated with the Abell clusters.

Figure 6.2. The central region of the Coma cluster of galaxies—an x-ray emitting cluster.

6.4.1. The Perseus Cluster

The Perseus cluster is the brightest and best studied of the x-ray cluster sources, and the only one which, at present, merits a separate description. The x-ray source was first reported by Fritz *et al* (1971) from a rocket observation, and its extended structure was noticed by Forman *et al* (1972) using data from the Uhuru satellite. The Uhuru observers found the x-ray source to have a diameter of approximately 0·5°, on the assumption that it was a disc of uniform surface brightness. Subsequent, more detailed studies of the spatial structure of the x-ray source have been made by Fabian *et al* (1974), using the Copernicus satellite, and by Wolff *et al* (1975), using a one-dimensional focusing system on a sounding rocket. From these later studies, it emerges that the cluster emission is brighter towards the centre of the cluster, fading away smoothly with distance from the centre. The dominant and active galaxy NGC 1275 forms an additional point-like source amounting to some 20 or 30% of the whole cluster emission. The rocket observations suggest that the NGC 1275 source has a harder spectrum than the diffuse cluster emission.

The best spectral data have been obtained using the Ariel

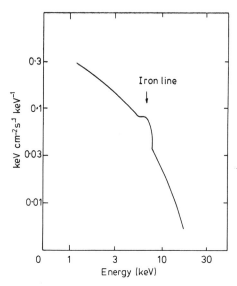

Figure 6.3. The x-ray spectrum of the Perseus cluster of galaxies. Data from Mitchell *et al* (1976).

130

V satellite. Mitchell *et al* (1976) obtained a spectrum of the whole cluster, including NGC 1275. This spectrum showed clear evidence of line emission at about 6·7 keV from highly ionised iron atoms, whilst the overall spectrum could be fitted with a composite thermal bremsstrahlung continuum, consistent with the adiabatic gas sphere model mentioned below. In the case of Perseus, there is therefore clear evidence that the cluster emission is thermal in origin. The possibility that NGC 1275 itself is a non-thermal source (as is suspected for other active galaxies) is quite consistent with these spectral measurements.

6.4.2. *The Mechanism of Cluster X-ray Emission*

Two mechanisms have been seriously considered to account for cluster x-ray emission: inverse Compton radiation from collisions between cosmic ray electrons and microwave background photons, and thermal emission from a tenuous gas at about 10^8 K. Synchrotron theories run into serious difficulties in explaining sources as large as the cluster regions (Brecher and Burbidge 1972). There now exists quite definite evidence in favour of the thermal mechanism. The Perseus cluster spectrum clearly shows the iron emission line which is characteristic of thermal emission, and spectral data on the Virgo, Coma and Centaurus clusters also favour thermal emission.

Thermal models for cluster emission have been constructed by Lea *et al* (1973) and by Gull and Northover (1975). The first model treats the emitting region as a sphere of gas at a single high temperature, and the latter model takes into account the likely temperature gradients in the sphere of gas, the centre being hottest. The Gull and Northover 'adiabatic gas sphere' model is consistent with the detailed spectrum of the Perseus cluster source. Both models permit the calculation of the mass of material involved in the x-ray emission. It is clear that this mass, added to the masses of the galaxies, is still insufficient to provide the gravitational binding force to hold the clusters together.

6.5. The Diffuse X-Ray Background

Unlike the night-time sky at visible wavelengths, the x-ray sky is bright. As well as containing the discrete sources with

131

which most of this book has been concerned, the whole sky exhibits a background surface brightness in x-rays. At energies greater than about 2 keV, this background is isotropic; this suggests that it must be extragalactic, rather than Galactic, in origin. The argument runs as follows. The Galaxy is essentially transparent to x-rays of energy greater than 2 keV, so that if a diffuse emission mechanism were distributed in the same way as the disc-shaped stellar population of the Galaxy, the diffuse flux would appear brighter in the plane of the Milky Way than at the Galactic poles. Very little extra diffuse emission is seen in the Milky Way (Cooke *et al* 1969, Bleach *et al* 1972), thus definitely ruling out the possibility of a Galactic origin for the background radiation. The scattering of solar x-rays in the Earth's atmosphere will certainly not account for the bright x-ray sky above 2 keV. At energies below 2 keV, the sky is still bright in x-rays, but the flux is stronger at the Galactic poles than in the plane of the Milky Way. It seems that both Galactic and extragalactic factors contribute, and the absorption by the interstellar gas in the Galaxy has to be taken into account. We may note that the only other region of the spectrum in which the night sky is bright is the microwave region.

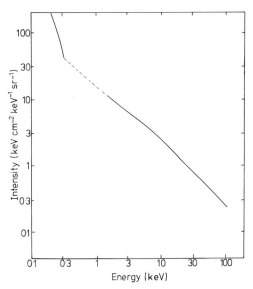

Figure 6.4. The spectrum of the isotropic background radiation.

6.5.1. The Isotropic Background $(E > 2\,keV)$

Only two types of measurement have been made on the diffuse x-ray background: the determination of its spectrum, and the determination of its degree of isotropy.

Spectral measurements are fairly difficult to make. It will be recalled that the normal technique for observing discrete sources is to scan across such a source with a directionally sensitive detector (or 'telescope'). In this way, both the detector response to the source and its response to blank sky are obtained. By subtracting the blank sky response from the source response, one obtains a signal which is characteristic of the source alone. There is no requirement for a detailed understanding of the blank sky response for this procedure to be valid. The blank sky response, however, is caused by two factors: one is the diffuse x-ray background of the sky, and the other is the response of the detector to cosmic ray particles, gamma-rays and any radioactivity in the materials of the instrument and spacecraft. When measuring the spectrum of the x-ray background, therefore, it is necessary to measure, and subtract out, the cosmic ray and radioactivity effects. Techniques for doing this include (i) placing a shutter over the entrance to the telescope, and (ii) pointing the telescope at the Earth. Neither technique is perfect, because both alter the geometry of the x-ray instrument with respect to the incoming cosmic ray flux, and yield readings which are not truly representative of the cosmic ray response when looking at the sky. Additional problems with satellite instruments are that they respond to geomagnetically trapped particles, which give rise to a variable response around the orbit, and can induce radioactivity in scintillation counter detectors.

The spectrum of the diffuse background falls monotonically with increasing x-ray energy and can be observed throughout the x-ray and gamma-ray regions. Measurements over a ten to one energy range are generally consistent with a power-law spectral form. There is some evidence that the spectrum is flatter below 10 keV than above, though the precise amount by which the spectral index changes is a matter of dispute between different authors (Dennis *et al* 1973, Adams

133

and Ricketts 1973, Schwartz and Petersen 1974). There appears to be no strong 6·7 keV line present in the spectrum (Boldt *et al* 1971) and, therefore, no spectral clue as to the mechanism generating the background.

Isotropy measurements are made by scanning regions of sky which are devoid of known discrete sources, and searching for systematic differences in the recorded fluxes. This again is difficult because great care must be exercised in eliminating changes due to other causes, such as the changing position of a satellite in its orbit. The most sensitive measurement of this type has been made by the Uhuru satellite using its $5 \times 5°$ field of view collimator. The results suggest that the x-ray flux from blank areas of sky varies by about 3% from one $5 \times 5°$ area to the next (Schwartz *et al* 1976) Pye and Warwick (1979) discuss Ariel V results.

There are two schools of thought on the origin of the x-ray background. Either (a) the background is truly diffuse, being generated more or less uniformly in space, or (b) the background is produced by a large number of discrete sources, which are so numerous as to give the appearance of a diffuse source, in the same way that, to the naked eye, the Milky Way appears diffuse, although it is really composed of individual stars. The isotropy measurements are consistent with either school of thought. If the background has a genuinely diffuse origin, then the small directional variations observed can be attributed to the presence of discrete sources which are too faint to have been identified and catalogued as such. On the other hand, if the background is really the superposition of discrete sources, the low value of the anisotropy implies that there must be a very large number of them contributing. Detailed analysis of the isotropy in terms of discrete source models is complex (see, for example, Fabian 1972, 1975). It is clear that the number of sources in the whole sky at distances closer than the distance at which objects show a redshift of 1, is in excess of 10^7.

Considering the size of the problem of accounting for the diffuse extragalactic background, rather little has been published on the subject. Silk (1973) reviewed the theories which do exist. Amongst the diffuse source theories, inverse Compton scattering of microwave photons by relativistic electrons

134

and the thermal emission of a hot intergalactic medium are possibilities. Since little is known about either the relativistic electron density, or the gaseous composition of the intergalactic medium, it is difficult to make further progress with these theories. Discrete source theories require sources with a high luminosity and a high space density to account for the observed flux. No known class of extragalactic x-ray is capable of accounting for the observed flux unless source evolution is invoked, although galaxy cluster sources and Seyfert galaxies might each account for 10% of the observed flux. Source evolution theories postulate that extragalactic x-ray emitters were more luminous in the past. The radiation we observe from very distant galaxies would have left them in the distant past, so we might in fact be seeing a strong contribution from objects at cosmological distances.

6.5.2. *The Ultra-soft X-ray Background*

At very soft x-ray energies below about 1 keV, the sky is bright in x-rays, but the radiation is no longer isotropic. Care must be taken when making measurements that the readings are not contaminated by solar x-rays scattered by the Earth's atmosphere. The flux at the Galactic poles lies well above the extrapolation of the higher-energy data, and the Galactic poles are appreciably brighter than the Galactic plane. The gas in the Galactic disc is expected to absorb ultra-soft x-rays and, at first, the observations were interpreted as indicating an ultra-soft x-ray background of extragalactic origin. There are, however, two major objections to this view. Firstly, the radiation in the plane of the Galaxy is brighter than would be expected for an absorbed extragalactic flux. Secondly, ultra-soft x-radiation of extragalactic origin would be absorbed by the Magellanic Clouds, and by M31, leaving dark shadows in the bright x-ray sky. These shadows have been looked for by McCammon *et al* (1971) for the SMC, by Rappaport *et al* (1975) for the LMC and by Margon *et al* (1974) for M31. No dark shadows have been observed. One is forced to conclude that an appreciable fraction of the ultra-soft background is of local Galactic production. The brightening towards the

Galactic poles can be understood if the source region in the Galactic disc is greater in Galactic latitudes than the distribution of the absorbing gas (Gorenstein and Tucker 1972). There is structure in the ultra-soft background which probably corresponds with the remains of very old supernova remnants. A map of the ultra-soft background can be found in De Korte *et al* (1974). Spectral details are reported by Levine *et al* (1976).

References

This reference list has been provided to enable the serious reader to follow up the text in the original literature. This list is far from complete, but should provide enough information for the reader to be able to find other references on a particular topic. Earlier references on a topic are usually listed at the end of a paper. Later references can be found by using the *Science Citation Index*.

References are listed in abbreviated form. Some journals have special sections with different page numbers—an *Astrophysical Journal* reference with a page number preceded by an 'L' means that the article is to be found in the Letters section of the journal. Similarly, a 'P' in a *Monthly Notices of the Royal Astronomical Society* reference means the pink pages section of the journal. Only one author's name is given. No slight is intended to the co-authors. In any case, the number of occurrences of a particular author's name may not fairly represent his or her contribution to the subject.

Chapter 1

Culhane 1977 *Vistas in Astronomy* ed A Beer (New York: Pergamon) (reviews x-radiation from the Sun, a topic not covered in this volume)
Giacconi *et al* 1962 *Phys. Rev. Lett.* **9** 439 (the discovery paper)

Chapter 2

Adams *et al* 1972 *Astron. Astrophys.* **20** 121
Giacconi *et al* 1969 *Space Sci. Rev.* **9** 3
Gorenstein *et al* 1968 *Astrophys. J.* **153** 885
Gursky *et al* 1966 *Astrophys. J.* **146** 310
Schnopper *et al* 1968 *Space Sci. Rev.* **8** 534
—— 1970 *Astrophys. J.* **161** L161

(NB. Instrumentation is very poorly covered in the astronomical literature. More details can sometimes be found in journals such as the *Proceedings of the IEEE*.)

Chapter 3

Allen 1963 *Astrophysical Quantities* (London: Athlone Press)
Brown and Gould 1970 *Phys. Rev.* **D1** 2252
Ginzburg and Syrovatskij 1963 *The origin of Cosmic Rays* (New York: Pergamon)
Rose 1973 *Astrophysics* (New York: Holt, Rinehart and Winston)
Tucker and Koren 1971 *Astrophys. J.* **168** 283

Chapter 4

Centaurus X-3

Baity *et al* 1974 *Astrophys. J.* **187** 341
Cooke and Pounds 1971 *Nature Phys. Sci.* **229** 144
Giacconi *et al* 1971 *Astrophys. J.* **167** L67
Krzeminski 1974 *Astrophys. J.* **192** L135
Parkinson *et al* 1974 *Nature* **249** 746
Pounds *et al* 1975 *Mon. Not. R. Astron. Soc.* **173** 473
Schreier *et al* 1972 *Astrophys. J.* **172** L79

Hercules X-1

Bahcall and Bahcall 1972 *Astrophys. J.* **178** L1
Bopp *et al* 1973 *Astrophys. J.* **186** L123
Clark *et al* 1972 *Astrophys. J.* **177** L109
Crampton 1974 *Astrophys. J.* **187** 345
Crampton and Hutchings 1972 *Astrophys. J.* **178** L65
Davidsen *et al* 1972 *Astrophys. J.* **177** L97
Doxsey *et al* 1973 *Astrophys. J.* **182** L25
Forman *et al* 1972 *Astrophys. J.* **177** L103
Giacconi *et al* 1973 *Astrophys. J.* **184** 227
Holt *et al* 1976 *Nature* **263** 484
Jones *et al* 1973 *Astrophys. J.* **182** L109
Middleditch and Nelson 1973 *Astrophys. Lett.* **14** 129
Pravdo *et al* 1977 *Astrophys. J.* **215** L61
Tananbaum *et al* 1972 *Astrophys. J.* **174** L143
Trumper *et al* 1978 *Astrophys. J.* **219** L105
Ulmer *et al* 1972 *Astrophys. J.* **178** L61

Scorpio X-1

Ables 1969 *Astrophys. J.* **155** L27
Acton *et al* 1970 *Astrophys. J.* **161** L175
Cowley and Crampton 1975 *Astrophys. J.* **201** L65
Giacconi *et al* 1962 *Phys. Rev. Lett.* **9** 439
Gorenstein *et al* 1968 *Astrophys. J.* **153** 885
Gottlieb *et al* 1975 *Astrophys. J.* **195** L33
Grader *et al* 1970 *Astrophys. J.* **159** 201
Griffiths *et al* 1971 *Nature Phys. Sci.* **229** 175
Gursky *et al* 1966a *Astrophys. J.* **144** 1249
—— 1966b *Astrophys. J.* **146** 310

Hiltner and Mook 1970 *Ann. Rev. Astron. Astrophys.* **8** 139
Holt *et al* 1969 *Astrophys. J.* **158** L155
Lewin *et al* 1968 *Astrophys. J.* **152** L55
Sandage *et al* 1966 *Astrophys. J.* **146** 316
Stockman *et al* 1973 *Astrophys. J.* **183** L63
Tucker and Koren 1971 *Astrophys. J.* **168** 283
Wallerstein 1967 *Astrophys. Lett.* **1** 31
Westphal *et al* 1968 *Astrophys. J.* **154** 139
White *et al* 1976 *Mon. Not. R. Astron. Soc.* **176** 91

Cygnus X-1

Agrawal *et al* 1972 *Astrophys. Space Sci.* **18** 408
Bahcall *et al* 1974 *Astrophys. J.* **189** L17
Braes and Miley 1971 *Nature* **232** 246
—— 1972 *Nature Phys. Sci.* **235** 147
Bregman *et al* 1973 *Astrophys. J.* **185** L117
Bolton 1972 *Nature Phys. Sci.* **240** 124
Brucato and Kristian 1973 *Astrophys. J.* **179** L129
Byram *et al* 1966 *Science* **152** 66
Giacconi *et al* 1967a *Astrophys. J.* **148** L119
—— 1967b *Astrophys. J.* **148** L129
Haymes *et al* 1968 *Astrophys. J.* **151** L125
Hjellming 1973 *Astrophys. J.* **182** L29
Hjellming and Wade 1971 *Astrophys. J.* **168** L21
Holt *et al* 1976 *Astrophys. J.* **203** L63
Hutchings *et al* 1973 *Astrophys. J.* **182** 549
Lester *et al* 1973 *Nature Phys. Sci.* **241** 125
McCracken *et al* 1966 *Science* **154** 1000
Margon *et al* 1973 *Astrophys. J.* **185** L113
Murdin and Webster 1971 *Nature* **233** 110
Oda *et al* 1971 *Astrophys. J.* **166** L1
Rappaport *et al* 1971 *Astrophys. J.* **168** L17
Rothschild *et al* 1974 *Astrophys. J.* **189** L13
Sanford *et al* 1974 *Astrophys. J.* **190** L55
Schreier *et al* 1971 *Astrophys. J.* **170** L21
Tananbaum *et al* 1972 *Astrophys. J.* **177** L5
Trimble *et al* 1973 *Mon. Not. R. Astron. Soc.* **162** 1
Wade and Hjellming 1972 *Nature* **235** 271
Webster and Murdin 1972 *Nature* **235** 37

Cygnus X-3

Becklin *et al* 1972 *Nature Phys. Sci.* **239** 130
—— 1973 *Nature* **245** 302
Bleach *et al* 1972 *Astrophys. J.* **171** 51
Braes and Miley 1973 *Nature* **237** 506
Canizares *et al* 1973 *Nature Phys. Sci.* **28**
Chu and Beiging 1973 *Astrophys. J.* **179** L22

Conner *et al* 1972 *Nature Phys. Sci.* **239** 125
Dashido *et al* 1974 *Nature* **251** 36
Giacconi *et al* 1967 *Astrophys. J.* **148** L119
Gregory *et al* 1972 *Nature* **239** 440
Hjellming 1973 *Science* **182** 1089
Holt *et al* 1976 *Nature* **260** 593
Kestenbaum *et al* 1977 *Astrophys. J.* **216** L19
Lamb *et al* 1977 *Astrophys. J.* **212** L63
Parsignault *et al* 1972 *Nature Phys. Sci.* **239** 123
Sanford and Hawkins 1972 *Nature Phys. Sci.* **239** 135
Serlemitsos *et al* 1975 *Astrophys. J.* **201** L9

X-ray Transients
Boley *et al* 1976 *Astrophys. J.* **203** L13
Chodil *et al* 1967 *Phys. Rev. Lett.* **19** 681
Davis *et al* 1975 *Nature* **257** 659
Doxsey *et al* 1976 *Astrophys. J.* **203** L9
Eachus *et al* 1976 *Astrophys. J.* **203** L17
Elvis *et al* 1975 *Nature* **257** 656
Gull *et al* 1976 *Astrophys. J.* **206** 260
Hoffman *et al* 1976 *Nature* **261** 208
Kaluzienski *et al* 1975 *Nature* **256** 633
—— 1977 *Astrophys. J.* **212** 203
Matilsky *et al* 1976 *Astrophys. J.* **210** L127
Owen *et al* 1976 *Astrophys. J.* **203** L15
Rappaport *et al* 1976 *Astrophys. J.* **208** L119
Ricketts *et al* 1975a *Nature* **256** 631
—— 1975b *Nature* **257** 657
Rosenberg *et al* 1975 *Nature* **256** 628
Tsunemi *et al* 1977 *Astrophys. J.* **211** L15
Wickramasinghe and Warren 1976 *Mon. Not. R. Astron. Soc.* **177** 59P

Galactic Bulge Sources
Davison and Morrison 1977 *Mon. Not. R. Astron. Soc.* **178** 53P
Janes *et al* 1972 *Nature* **235** 152
Wilson *et al* 1977 *Astrophys. J.* **215** L111

Globular Cluster Sources
Bahcall and Ostriker 1975 *Nature* **256** 23
Clark *et al* 1975 *Astrophys. J.* **199** L93
Ulmer *et al* 1976 *Astrophys. J.* **208** 47

Bursters
Carpenter *et al* 1976 *Nature* **262** 473
Clark *et al* 1976 *Astrophys. J.* **207** L105
Forman and Jones 1976 *Astrophys. J.* **207** L177
Grindlay and Gursky 1976 *Astrophys. J.* **205** L127
Hoffman *et al* 1976 *Nature* **261** 208

Lewin and Joss 1977 *Nature* **270** 211
Lewin *et al* 1976 *Astrophys. J.* **207** L95
Liller 1977 *Astrophys. J.* **213** L21

Chapter 5

Cassiopeia A
Charles *et al* 1977 *Mon. Not. R. Astron. Soc.* **178** 307
Davison *et al* 1976 *Astrophys. J.* **206** L37
Fabian *et al* 1973 *Nature Phys. Sci.* **242** 18
Friedman *et al* 1967 *Science* **156** 374
Giacconi *et al* 1972 *Astrophys. J.* **178** 281
Gorenstein *et al* 1970 *Astrophys. J.* **160** 947
Hagen 1955 *Astrophys. J.* **122** 361
Hogg 1969 *Astrophys. J.* **155** 1099
Holt *et al* 1973
Mayer *et al* 1965 *Astrophys. J.* **141** 867
Pravdo *et al* 1976 *Astrophys. J.* **206** L41
Rosenberg *et al* 1970 *Mon. Not. R. Astron. Soc.* **147** 215
Van den Bergh 1971 *Astrophys. J.* **165** 457
Van den Bergh and Dodd 1970 *Astrophys. J.* **162** 485
Woltjer 1972 *IAU Symp. Proc., Uppsala* (Dordrecht: Reidel) p 277
—— 1974 *Lecce Conf. Proc.* (Dordrecht: Reidel) p 323

Puppis A
Baade and Minkowski 1955 *Astrophys. J.* **119** 206
Burginyon *et al* 1975 *Astrophys. J.* **200** 163
Catura and Acton 1976 *Astrophys. J.* **207** L163
Moore and Garmire 1976 *Astrophys. J.* **206** 247
Palmieri *et al* 1971 *Astrophys. J.* **164** 61
Seward *et al* 1971 *Astrophys. J.* **169** 515
Tucker and Koren 1971 *Astrophys. J.* **168** 283
Zarnecki and Culhane 1977 *Mon. Not. R. Astron. Soc.* **178** 57P
Zarnecki *et al* 1973 *Nature Phys. Sci.* **243** 4

Vela X
Brandt *et al* 1976 *Astrophys. J.* **208** 109
Clark and Culhane 1976 *Mon. Not. R. Astron. Soc.* **175** 573
Milne 1968 *Austral. J. Phys.* **21** 201
Moore and Garmire 1975 *Astrophys. J.* **199** 680
—— 1976 *Astrophys. J.* **206** 247
Thomson *et al* 1977 *Astrophys. J.* **214** L17
Wallace *et al* 1977 *Nature* **266** 692

Crab Nebula
Bowyer *et al* 1964 *Science* **146** 912
Charles and Culhane 1977 *Astrophys. J.* **211** L23
Davison *et al* 1975 *Nature* **253** 610

Fazio *et al* 1971 *Proc. 11th Int. Conf. on Cosmic Rays* (Budapest: Akademiai Kiado) p 115
Fritz *et al* 1969 *Science* **164** 709
—— 1971 *Astrophys. J.* **164** L55
Hillier *et al* 1970
Ku *et al* 1976 *Astrophys. J.* **204** L77
Novick *et al* 1972 *Astrophys. J.* **174** L1
Parlier *et al* 1973 *Nature Phys. Sci.* **242** 117
—— 1974 *Lecce Conf. Proc.* in *Astrophys. Space Sci. Library* **45** 267
Palmieri *et al* 1975 *Astrophys. J.* **202** 494
Rappaport *et al* 1971 *Nature Phys. Sci.* **229** 40
Ricker *et al* 1975 *Astrophys. J.* **197** L83
Scargle 1969 *Astrophys. J.* **156** 401
Staelin and Reifenstein 1968 *Science* **162** 1481
Staubert *et al* 1975 *Astrophys. J.* **201** L15
Toor and Seward 1974 *Astron. J.* **79** 995
Toor *et al* 1976 *Astrophys. J.* **207** 96
Weisskopf *et al* 1978 *Astrophys. J.* **220** L117
Wolff *et al* 1975 *Astrophys. J.* **202** L15

Chapter 6

Margon and Ostriker 1973 *Astrophys. J.* **186** 91

Normal galaxies
Bowyer *et al* 1974 *Astrophys. J.* **190** 285
Liller 1973 *Astrophys. J.* **184** L37
Schreier *et al* 1972 *Astrophys. J.* **178** L71
Tuohy and Rapley 1975 *Astrophys. J.* **198** L69

Active galaxies
Auriemma *et al* 1978 *Astrophys. J.* **221** L7
Becklin *et al* 1971 *Astrophys. J.* **170** L15
Bowyer *et al* 1970 *Astrophys. J.* **161** L1
Catura *et al* 1972 *Astrophys. J.* **177** L1
Cooke *et al* 1978a *Mon. Not. R. Astron. Soc.* **182** 489
—— 1978b *Mon. Not. R. Astron. Soc.* **182** 661
Culhane 1978 *Q. J. R. Astron. Soc.* **19** 1
Davison *et al* 1975 *Astrophys. J.* **196** L23
Elvis *et al* 1978 *Mon. Not. R. Astron. Soc.* **183** 129
Grindlay *et al* 1975 *Astrophys. J.* **201** L133
Gursky *et al* 1971 *Astrophys. J.* **165** L43
Ives *et al* 1976 *Astrophys. J.* **207** L159
Lawrence *et al* 1977 *Mon. Not. R. Astron. Soc.* **181** 93P
Price and Stull 1975 *Nature* **255** 467
Stark *et al* 1976 *Mon. Not. R. Astron. Soc.* **174** 35P
Winkler and White 1975 *Astrophys. J.* **199** L139
Wolff *et al* 1976 *Astrophys. J.* **208** 1

Clusters of Galaxies

Brecher and Burbidge 1972 *Nature* **237** 440
Culhane 1978 *Q. J. R. Astron. Soc.* **19** 1
Fabian *et al* 1974 *Astrophys. J.* **189** L59
Forman *et al* 1972 *Astrophys. J.* **178** 309
Fritz *et al* 1971 *Astrophys. J.* **164** L81
Gorenstein and Tucker 1977 *Ann. Rev. Astron. Astrophys.* **14** 373
Gull and Northover 1975 *Mon. Not. R. Astron. Soc.* **173** 585
Gursky *et al* 1971 *Astrophys. J.* **167** L81
Kellogg *et al* 1973 *Astrophys. J.* **185** L13
Lea *et al* 1973 *Astrophys. J.* **184** 105
Mitchell *et al* 1976 *Mon. Not. R. Astron. Soc.* **175** 29P
Ryle and Windram 1968 *Mon. Not. R. Astron. Soc.* **138** 1
Welch and Sastry 1971 *Astrophys. J.* **169** L3
Wolff *et al* 1975 *Astrophys. J.* **193** L53

Background

Adams and Ricketts 1973 *Astrophys. Space Sci.* **24** 585
Bleach *et al* 1972 *Astrophys. J.* **164** L101
Boldt *et al* 1971 *Astrophys. J.* **167** L1
Cooke *et al* 1969 *Nature* **224** 134
Dennis *et al* 1973 *Astrophys. J.* **186** 97
Fabian 1972 *Nature Phys. Sci.* **237** 19
—— 1975 *Mon. Not. R. Astron. Soc.* **172** 149
Gorenstein and Tucker 1972 *Astrophys. J.* **176** 333
De Korte *et al* 1974 *Astrophys. J.* **190** L5
Levine *et al* 1976 *Astrophys. J.* **205** 226
McCammon *et al* 1971 *Astrophys. J.* **168** L33
Margon *et al* 1974 *Astrophys. J.* **191** L117
Pye and Warwick 1979 *Mon. Not. R. Astron. Soc.*
Rappaport *et al* 1975 *Astrophys. J.* **196** L15
Schwartz and Petersen 1974 *Astrophys. J.* **190** 297
Schwartz *et al* 1976 *Astrophys. J.* **204** 315
Silk 1973 *Ann. Rev. Astron. Astrophys.* **11** 269

Appendix

Relative strengths of some of the x-ray sources mentioned in the text. Uhuru catalogue names and strengths are quoted.

Name	4U Catalogue Number	Strength	Comments
Cen X-3	1118 −60	200	Eclipsing binary
Her X-1	1656 +35	100	Eclipsing binary
Vela X-1	0900 −40	250	Eclipsing binary
Sco X-1	1617 −15	17000	Brightest source
Cyg X-1	1956 +35	1175	Black hole binary
Cir X-1	1516 −56	720	Black hole binary?
Cyg X-3	2030 +40	385	Gamma-ray source
SMC X-1	0115 −73	36	Extragalactic eclipsing binary
A0620 −00	—	70000	Approx. at maximum (transient)
GX3 +1	1744 −26	600	Bright Galactic bulge source
GX9 +1	1758 −20	600	Bright Galactic bulge source
NGC 1851	0513 −40	18	Globular cluster
Cas A	2321 +58	53	Young supernova remnant
Tycho	0022 +63	9	Young supernova remnant
Crab	0531 +21	947	Standard source
M31	0037 +39	2·4	Nearby normal galaxy
Cen A	1322 −42	8	Radio galaxy
Cyg A	1959 +40	4	Radio galaxy
NGC 4151	1206 +39	43	Nearby Seyfert
3C 273	1226 +02	2·7	Quasar
Perseus	0316 +41	47	Cluster + NGC 1275
Virgo	1228 +12	22	Cluster + M87 + galaxies
Abell 2256	1707 +78	3·6	Cluster of galaxies

Notes

Uhuru strengths relate to the 2–6 keV region and are explained in Chapter 1.

The older supernova remnants have been excluded from this list as they emit the majority of their x-rays below Uhuru energies.

4U catalogue members have been included because they indicate the approximate right ascension and declination of the source.

Variable sources have their maximum strengths quoted.

Index